If you mean that the proximity of one color should give beauty to another that terminates near it, observe the rays of the sun in the composition of the rain- bow, the colors of which are generated by the falling rain, when each drop in its descent takes every color of the bow.

—Leonardo da Vinci, *Treatise on Painting*, 1490s

The Book of
RAINBOWS

Art, Literature, Science, & Mythology

By Richard Whelan
Designed by Arnold Skolnick

FIRST GLANCE BOOKS, COBB, CALIFORNIA

Frontispiece
ERNST HASS, *Rainbow, Equador*, N.D.

ACKNOWLEDGMENTS

I wish to extend warm thanks to all of the many people who, in one way or another, aided in the preparation of this book. I must give very special thanks to my friend Jack Faxon, who worked a couple of miracles in my behalf. My friend the photographer David Mosconi inspired me to begin work on the book, and the enthusiasm of Oceanic Graphic Printing's David Li and of First Glance's Rodney Grisso and Neil Panico made it possible to transform the idea into reality.

I wish to acknowledge my immense debt of gratitude to the New York Public Library, especially to the staff in the main reading room and in the Miriam & Ira D. Wallach Division of Art, Prints, and Photographs. I also thank the art library and the Nielson Library of Smith College, where I received a gracious welcome and exceptionally helpful assistance.

Chameleon Books and First Glance Books join me in thanking all of the following for their kind assistance: The George Adams Gallery; Jorge Amat; Art and Commerce; Tom Bean; John Berggruen Gallery; Judy Bond at the Cline Fine Art Gallery; James Crawford at the Canajoharie Library and Art Gallery; Verna Curtis at the Library of Congress; Margaret Daly at the National Gallery, London; Courtney DeAngelis at the Amon Carter Museum; Darlene Dueck at the Anschutz Collection; France Duhamel at the National Gallery of Canada; Morris Engel; Melody Ennis at the Museum of Art, Rhode Island School of Design; Janet Fish; Hanni Forester at the Asian Art Museum of San Francisco; Beth Garfield at the Detroit Institute of Arts; Marie-Thérèse Gousset at the Bibliothèque Nationale de France; Douglas Gruenau; Mary Haas at the Fine Arts Museums of San Francisco; Lisa Hintzpeter at the Brooklyn Museum of Art; Marilyn Hunt at the Yale Center for British Art; Lisa Jacobson at the Minnesota Museum of American Art; Alison Jasonides at Art Resource; Tom Kennedy at the A:shiwi A:wan Museum and Cultural Center, Zuni, New Mexico; Joel and Kate Kopp of America Hurrah; Vivian Kurz; Ellen Kutcher at Reynolda House, Museum of American Art, Winston-Salem; Jessica Mantaro at the Berry-Hill Galleries; Jo-Anne Mehring at the Robert McLaughlin Gallery, Oshawa, Ontario; Achim Moeller; Moke Mokotoff of Mokotoff Asian Arts; D.C. Moore Gallery; Anna Obolensky at Agence de Presse News Archives (A.N.A.), Paris; Tish O'Connor at Perpetua Press; Louisa Orto at Art Resource; Bebe Overmiller at the Library of Congress; Vincent Parrella; Yancey Richardson Gallery; Philip Rieser at the Ernst Haas Studio; Galen and Barbara Rowell of Mountain Light; Shelley and Donald Rubin; Eva Schorr at the Artists' Rights Society; Daniel Schulman at the Art Institute of Chicago; Helmut Selzer at the Historisches Museum der Stadt Wien; Fred Stern; Joan Tafoya at the Museum of Fine Arts, Museum of New Mexico, Santa Fe; Karen Tates at the Brooklyn Museum of Art; A. Alfred Taubman; Julia Van Haaften, Curator of Photographs at the New York Public Library; Faye Van Horne at the Art Gallery of Ontario; Paul Vucetich; Roberta Waddell, Curator of Prints at the New York Public Library; Anne Walker at Sotheby's; Susanne Waugh at Sotheby's; Amber Woods at the Wadsworth Atheneum, Hartford. We hope that anyone who has been inadvertently omitted from this list will accept our apologies.

Over the past fifteen years, Arnold Skolnick of Chameleon Books has designed several books of Robert and Cornell Capa's photographs that Cornell and I edited. The present book represents a new venture for Arnold and myself as full collaborators on a new series of visual-arts books.

Arnold's assistant at Chameleon Books, Laura MacKay, did a splendid job of coordinating the enormous task of obtaining transparencies and permissions, as well as providing invaluable help with other aspects of the project.

Thanks also to K.C. Scott and Lisa Carta for their technical assistance.

CONTENTS

The Rainbow in the Visual Arts

BECAUSE A RAINBOW always seems like a magical apparition, nearly everyone has indelible memories of spectacular ones or of ones seen under extraordinary circum-
stances. My own most vivid rainbow memory dates to
September 1990, when I was just beginning to work on a biog-
raphy of Alfred Stieglitz, who had been married to Georgia
O'Keeffe. Driving from northern New Mexico down to Abiquiu to
visit O'Keeffe's former house and studio, I rounded a curve in the
road at the crest of a hill near Abiquiu and suddenly saw a magnificent valley spread out before me. In it was a nearly perfect
double rainbow, which I gratefully took as an auspicious omen
for my project. Oddly, O'Keeffe herself seems never to have
painted a rainbow.

IT IS HARDLY SURPRISING that a phenomenon as beautiful,
poetic, and mysterious as the rainbow should have appeared
in many works of art from ancient times to the present day.
In medieval European art depictions of rainbows were
almost entirely limited to two Biblical scenes. The first was the
Old Testament story of how God told Noah after the terrible
Deluge, "I do set my bow in the cloud, and it shall be for a token
of a covenant between me and the earth. And it shall come to
pass, when I bring a cloud over the earth, that the bow shall be
seen in the cloud, and I will remember my covenant, which is
between me and you and every living creature of all flesh: and the
waters shall no more become a flood to destroy all flesh." The
second scene was the Last Judgment, in Flemish medieval and
Renaissance paintings of which Christ is usually enthroned on a
large circular rainbow as the virtuous are admitted into heaven
and the sinners are cast down into hell. This iconography derives
from the following passage in the biblical book of Revelation
(chapter 4, verses 1-3):

> After this I looked, and lo, in heaven an open door! And the first
> voice, which I had heard speaking to me like a trumpet, said, "Come
> up hither, and I will show you what must take place after this." At
> once I was in the Spirit, and lo, a throne stood in heaven, with one
> seated on the throne! And he who sat there appeared like jasper and
> carnelian, and round the throne was a rainbow that looked like an
> emerald.

EVRARD D'ESPINGUES, *THE RAINBOW*, 1480

6

Noah and His Family Leaving the Ark, 14th century Mosaic in the Basilica of San Marco, Venice

7

JOHN MARTIN (1789–1854), *NOAH'S SACRIFICE AFTER THE FLOOD*, N.D.

PAOLO UCCELLO, *NOAH'S SACRIFICE AND NOAH'S DRUNKENNESS*, 1460s

Another passage in Revelation (10:1) states, "Then I saw another mighty angel coming down from heaven, wrapped in a cloud, with a rainbow over his head, and his face was like the sun, and his legs like pillars of fire."

In the book of Ezekiel (1:28), the prophet says of his vision of God, "Like the appearance of the bow that is in the cloud on the day of rain, so was the appearance of such brightness round about."

Representations of both biblical scenes are to be found in medieval mosaics, frescoes, and illuminated manuscripts, as well as in early Renaissance paintings and woodcuts. During the Renaissance, rainbows also began to appear occasionally in paintings of the Madonna and Child, as in Grünewald's *Stuppach Madonna* or in Raphael's *Madonna di Foligno*, perhaps in allusion to the traditional symbolism mentioned by the eighteenth-century American Calvinist minister Cotton Mather:

> The Covenant with our Father Noah, whereof we have the Rainbow for an Obsignation, had such an Aspect upon the Messiah that we may fairly be led by the Rainbow to remember the whole Covenant of Grace, in all the Very Great and Precious Promises of it.

With the revival of interest in ancient Greek and Roman literature in the fifteenth and sixteenth centuries came paintings of mythological scenes in which rainbows figured. For example, a golden rainbow arcs in the background of two of Titian's sensuous paintings of Venus and Adonis. In the baroque style of seventeenth-century France we find a rainbow illuminating the nude bodies of gods and goddesses in Jacques Stella's rather saccharine *Judgment of Paris*. But by far the most spectacular Olympian rainbow is that which arches across the ceiling of the throne room in the royal palace in Madrid, painted by Giambattista Tiepolo in the 1760s.

MATTHIAS GRÜNEWALD, *THE STUPPACH MADONNA*, 1517

10

Roger van der Weyden, *The Last Judgment*, 1450–51

SCHELTE ADAMS BOLSWERT, *LANDSCAPE WITH RAINBOW*, ENGRAVING AFTER RUBENS, 1640S

LANDSCAPE PAINTINGS—pure landscapes, unpopulated by saints, gods, or heroes—were an invention of the Protestant North, which condemned religious painting as a manifestation of Catholic idolatry and denounced mythological painting as a celebration of paganism. Edward Norgate, who was responsible for buying paintings for King Charles I of England, perhaps expressed the operative philosophy best when he explained that pure landscapes were "of all kinds of pictures the most innocent, and which the Divill himselfe could never accuse of idolatry."

Since generalities always invite contradiction, let it be said that it was the Flemish Catholic artist Peter Paul Rubens who first made something of a specialty of painting pure landscapes with rainbows. In addition to the splendid canvas reproduced in this book, Rubens painted a similar scene now in London's Wallace Collection, as well as others in the Louvre and the Hermitage. A rainbow also illuminates his *Landscape with the Shipwreck of St. Paul* in Berlin's Gemäldegalerie.

From then on, rainbows proliferated—as anyone may see by looking through the pictorial section of this book. An eye attuned to rainbows will soon begin to find them in many unexpected places. For instance, a rainbow can be seen through the window in Gilbert Stuart's standing portrait of George Washington (in the collection of the New York Public Library), presumably suggesting that American liberty, after the storm of the Revolution, is guaranteed by a new divine covenant. And in New Mexico, which Washington probably never imagined would become part of the nation he founded, we encounter a Native-American-style painting of a rainbow above the Spanish Colonial altar in the church of San Jose at the Laguna Pueblo.

Although proper treatment of the rainbow may seem to call for a fair-sized canvas, the phenomenon appealed to many eighteenth- and nineteenth-century artists working on a smaller scale in gouache or watercolor. Among the most notable of such works are examples by Paul Sandby (in Nottingham's Castle Museum, which also boasts Julius Caesar Ibbetson's canvas of Beeston Castle with a rainbow), by Thomas Girtin, Peter de Wint, and John Ruskin.

ANONYMOUS, *QUEEN ELIZABETH I WITH A RAINBOW*, C. 1580

13

JOHN CONSTABLE WROTE in his commentary for an engraving of one of his paintings, "Nature, in all the varied aspects of her beauty, exhibits no feature more lovely nor any that awakens a more soothing reflection than the rainbow, 'Mild arch of promise.'" Constable and J.M.W. Turner were rival British masters of the rainbow, but in nineteenth-century America George Inness was the undisputed holder of the title. Among his many rainbow landscapes, besides the two reproduced in this book, two that must be mentioned are *Scene in Perugia* (Boston Museum of Fine Arts) and *Delaware Water Gap* (Metropolitan Museum of Art, New York). In an article he wrote for *Harper's* in 1878 Inness declared, "Let Corot paint a rainbow, and his work reminds you of the poet's description, 'The rainbow is the spirit of the flowers.' Let Meissonier paint a rainbow, and his work reminds you of a definition in chemistry. The one is poetic truth, the other is scientific truth; the former is aesthetic, the latter is analytic." No one can doubt that Inness was on the side of Corot and poetic truth.

Although I have not succeeded in finding a rainbow by either Corot or Meissonier, the motif was popular in nineteenth-century France. The academic painter Jules Breton has left us a lovely oil sketch of a rainbow above a farm, not dissimilar to one by Jean Charles Cazin, whose view of a rainbow over a French village street is in the Boston Museum of Fine Arts. Among forerunners of the Barbizon School of landscape, Paul Huet painted *A View of the Valley and the Chateau d'Arques, near Dieppe*, and Georges Michel included a rainbow in his *Storm over the Seine Valley*. As for the Impressionists, those who depicted a rainbow at one time or another include Monet, Pissarro, and Manet, in whose Rubensesque costume-drama *La Pêche (Fishing at Saint-Ouen)* a rainbow makes its appearance in the background.

The rainbow's appeal to so-called visionary artists is obvious. William Blake leads the group with an especially splendid one in one of his illustrations for his poem *Jerusalem*, of which he embellished only one copy with watercolors and gold paint. Caspar David Friedrich's *Mountain Landscape with Rainbow* is also magnificent. Alas, a second Friedrich rainbow painting disappeared some time ago. The German architect and painter Karl Friedrich Schinkel seems to have had a great fondness for rainbows: they appear not only in his view of an imaginary cathedral but also in a sketch for an equestrian monument and in a painting of Zeus on Mount Olympus. Perhaps the British Pre-Raphaelites may be included among the visionaries, albeit rather bourgeois ones. Besides the paintings by John Everett Millais and Ford Madox Brown reproduced in this book, their depictions of rainbows include the first version of William Holman Hunt's *The Scapegoat*. The great Belgian visionary James Ensor painted a late self-portrait with a rainbow as well as an early seascape in which the subtlest of rainbows marks the end of a storm.

In the twentieth century, Robert and Sonia Delaunay's early abstractions look like shattered rainbows, but not until relatively late in his career did Robert get around to painting a reassembled one. Rainbows appear in works by artists as diverse as John Sloan, Marc Chagall, Jean Dufy, Otto Dix, Leonora Carrington, Jackson Pollock, Saul Steinberg, and H. C. Westermann. In the great contemporary collection of Cologne's Museum Ludwig we find not only Roy Lichtenstein's droll parody, *Landscape with Figures and Rainbow*, but also James Rosenquist's even droller *Rainbow*, in which the bands of color are drips down a shingled wall from wet paintbrushes laid side by side on a windowsill. Perhaps the most successful contemporary work incorporating a rainbow is Jim Dine's *Studio Landscape*, in which various cups and glasses of paint, as well as several empty bottles, constitute the treasure at the end of the rainbow.

Rainbows are rare in Asian art, except in Tibetan painting, where they have a specific spiritual meaning. Perhaps Chinese and Japanese artists—who have traditionally preferred delicate monochrome landscapes softened with mist and gentle rain—have disdained the rainbow as too vulgarly spectacular. However, the great Japanese printmaker Hiroshige delighted in exaggerations and distortions intended to shock an audience accustomed to refined understatement. It is therefore no surprise to find a rainbow in one of his works. With his playful wit, he makes a visual pun with watermelon rinds, which inversely mirror the rainbow's shape and echo its bands of color.

On a trip to the Grand Canyon in 1911, Arnold Genthe made—with the Autochrome color plates that had become available four years earlier—what may very well be the earliest color photograph of a rainbow. Of his extraordinary experience, he wrote:

In the deep gorges of the canyon there was rich picture material. One day I went down to the bottom on foot, and after most of my plates had been used up I started back on the seven-mile climb to the top, when I was caught in a terrific thunderstorm. It was not just one thunderstorm but a convention of them. As there was no place to seek shelter, all I could do was to trudge on carrying my camera, which seemed to be getting heavier every minute. When I got to the rim of the canyon, the storm had subsided, and there before me, descending into the purple depths, was the most magnificent rainbow I have ever beheld. I had just two plates left. Quickly I set up my camera. It was still drizzling and I had nothing to protect the lens from moisture, since my hat had blown off in the storm. Glancing around I saw a forest ranger on horseback to whom I signaled. He came cantering up. "What's the trouble?" he asked.

"I want to make a photograph of this rainbow," I said. "Would you be kind enough to hold your sombrero over the lens?"

Silently he did as I requested. When I had finished he gave me a doubtful look, asking, "Can you really take a picture of a rainbow?"

ARNOLD GENTHE, *RAINBOW IN THE GRAND CANYON*, 1911

HOKUSAI (1760–1849), *CUCKOO AND RAINBOW*, N.D.

17

OZENS OF OUTSTANDING PHOTOGRAPHERS, ranging from Laura Gilpin to Len Jenshel, have naturally recorded rainbows in color. But three of the greatest American photographers of the twentieth century—Alfred Stieglitz, Edward Weston, and Ansel Adams—photographed rainbows in black-and-white. In 1920 Stieglitz made a superb image of a rainbow over Lake George, where his family had their summer home. Weston photographed a rainbow from a mountaintop overlooking the Owens Valley of California one day in 1937 when he was driving with Ansel Adams. Weston's wife recounted the scene: "Stop the car! Out with the cameras! for a race between the photographers and the elements. Every second the light is shifting as the cloud-shadow patterns go racing by down in the valley. At one moment the Alabama Hills are a sooty menace of silhouette; at the next they are dazzling in a downpour of blinding light. Ansel and Edward work feverishly, stopping down, setting shutters, pulling slides, making negatives in split seconds. Rain sweeps over our mountain perch. Between negatives the photographers wring out their focusing cloths like wet towels." Adams went on to make his most memorable rainbow photograph—*Buddhist Grave Markers and Rainbow*—on the Hawaiian island of Maui in the mid-1950s.

AR BEYOND THE BOUNDARIES of the fine arts—or even of the decorative arts—exists a vast world of popular depictions of the rainbow. There are rainbows galore in children's books and on posters. Rainbows occur on orange-crate labels and on record jackets, on cereal boxes and on the sides of trucks, on calendars, greeting cards, sweaters, and so on ad infinitum. Neon rainbows have become a store-window cliché, and the producers of special effects can now easily brighten a dull movie or commercial with a digitized rainbow or two.

Although pop culture has threatened to trivialize the rainbow, a counter-current has elevated it to one of the most powerful and effective symbols of our time. I am, of course, thinking of the Rainbow Coalition, which champions universal tolerance and passionately advocates a world in which people of all colors, beliefs, and orientations will live and work together as harmoniously as all the colors cooperate to form a rainbow.

Middle Ages and
Renaissance

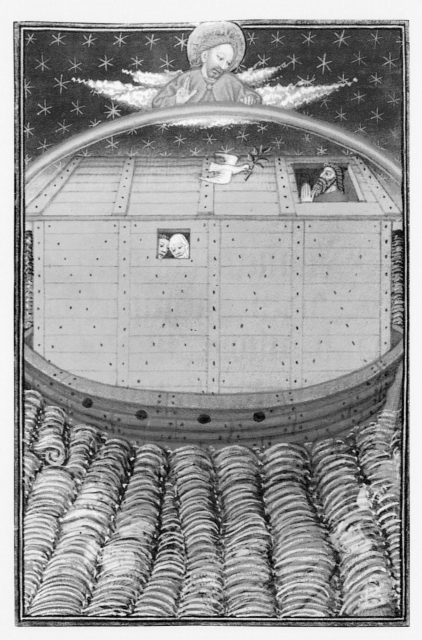

FASTOLF MASTER, *GOD ENDS THE DELUGE*, C. 1440–50

God's Pact with Noah, from the *Vienna Genesis*, 6th Century A.D

21

DOSSO DOSSI (GIOVANNI DE' LUTERI), *JUPITER, MERCURY, AND VIRTUE*, C. 1525

What skilful limner e'er would choose
To paint the rainbow's varying hues,
Unless to mortal it were given
To dip his brush in the dyes of Heaven?

—Sir Walter Scott, *Marmion, A Tale of Flodden Field,* 1808

Bernardino Pinturicchio,
Enea Silvio Piccolomini
Traveling to the Council of
Basel, 1505–07.
The rainbow refers to
Piccolomini and his retinue's
having survived a terrible
storm at sea.

As from the face of heaven the scatter'd clouds

Tumultuous rove, th'interminable sky

Sublimer swells, and o'er the World expands

A purer azure. Through the lightened air

A higher lustre and a clearer calm

Diffusive tremble; while, as if in sign

Of danger past, a glittering robe of joy,

Set off abundant by the yellow ray,

Invests the fields, and nature smiles revived.

—James Thomson, *The Seasons: Summer*, 1727

PETER PAUL RUBENS, *LANDSCAPE WITH RAINBOW*, AFTER 1635

Jacob Isaaksz van Ruisdael, *The Jewish Cemetery*, 1655–1660

He can behold
Things manifold
That have not been wholly told,—
Have not been wholly sung nor said.
For his thought, that never stops,
Follows the water-drops
Down to the graves of the dead,
Down through chasms and gulfs profound,
To the dreary fountain-head
Of lakes and rivers under ground;
And sees them, when the rain is done,
On the bridge of colors seven
Climbing up once more to heaven,
Opposite the setting sun.
Thus the Seer,
With vision clear,
Sees forms appear and disappear,
In the perpetual round of strange,
Mysterious change
From birth to death, from death to birth,
From earth to heaven, from heaven to earth;
Till glimpses more sublime
Of things unseen before,
Unto his wondering eyes reveal
The Universe, as an immeasurable wheel
Turning forevermore
In the rapid and rushing river of Time.

—Henry Wadsworth Longfellow (1807–1882)

The Science of Rainbows

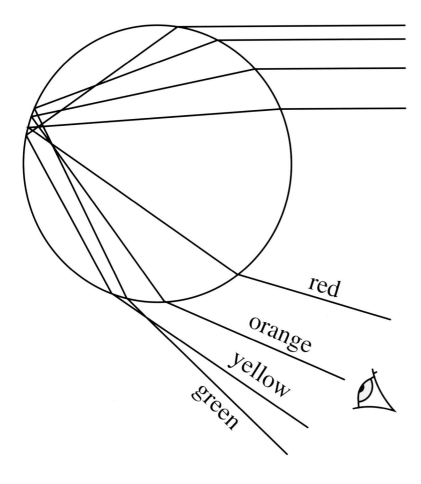

THE PATHS OF RAYS OF LIGHT IN A RAINDROP, SHOWING
HOW ONLY ONE COLOR FROM EACH DROP REACHES THE EYE

red

orange

yellow

green

STRICTLY SPEAKING, there is no such thing as a rainbow, for a rainbow is not something that has concrete existence occupying a specific space. Instead, a rainbow is an optical phenomenon so complex that each eye of any single observer receives light of a slightly differing wavelength from a given raindrop at any given moment—so that each eye actually sees a different rainbow. And, as each drop falls, the particular wavelength of light from it reaching the observer's eye changes constantly—so the rainbow seen by each eye also changes constantly. To complicate matters still further, the waves of light are not themselves colored; rather, waves of a particular length produce the sensation of their corresponding color by stimulating the optic nerves, which provide the necessary information to the brain.

Among the conditions necessary in order for a rainbow to occur, the most essential are, first, rain falling in the direction toward which the observer is facing and, second, bright sunlight (or, on exceedingly rare occasions, bright moonlight from a full moon) coming from a relatively low point above the horizon behind the observer.

A symmetrical rainbow in which the seven colors of the spectrum—red, orange, yellow, green, blue, indigo, and violet—are most perfectly articulated will be formed by relatively large raindrops (several millimeters in diameter); they must be quite spherical, not appreciably distorted by air resistance or wind. The drops that produce a rainbow are rarely more than two miles away from the observer and can sometimes be much closer. Someone recently told me of having seen a rainbow in Hawaii (which is known for the extraordinary frequency with which rainbows occur) one foot of which was on the ocean only about twenty-five feet from the spot on the beach where she was standing.

A rainbow most often follows, but may sometimes precede, precipitation. A thunderstorm is more likely than a gentle rain to provide the combination of large drops and abrupt onset or clearing needed to form a rainbow.

Each spherical raindrop acts as a tiny prism. Upon entering the upper part of a raindrop, sunlight is bent (refracted) slightly downward and broken into its spectrum of wavelengths. It is then reflected off the back of the drop. As it leaves the lower part of the raindrop, the light is again refracted, each wavelength to a distinctive extent, violet being bent the most and red the least.

30

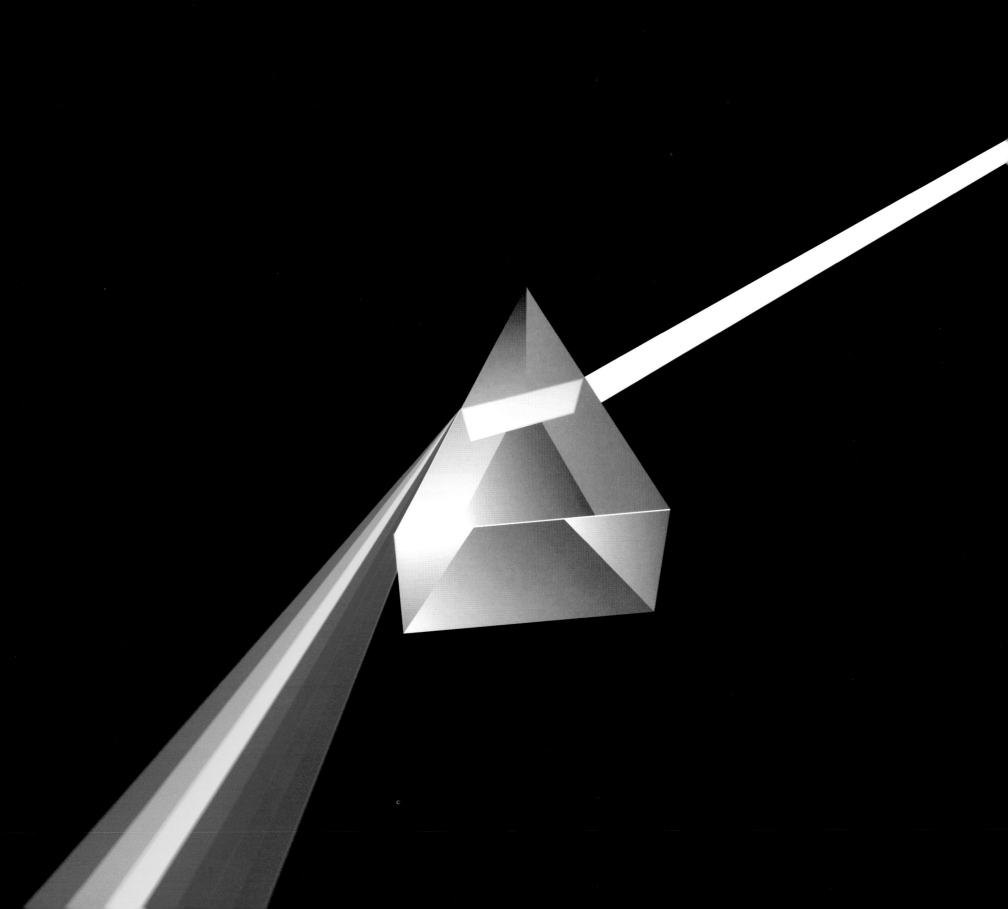

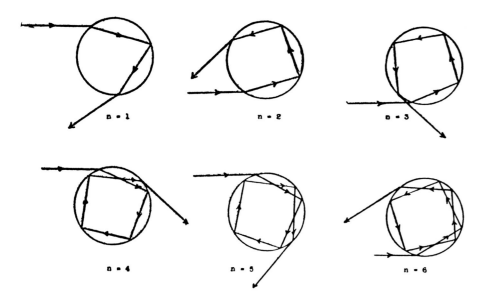

n = 1 n = 2 n = 3

n = 4 n = 5 n = 6

THE PATHS OF RAYS OF LIGHT WITHIN A RAINDROP TO PRODUCE
THE PRIMARY THROUGH THE SIXTH ORDER OF RAINBOWS

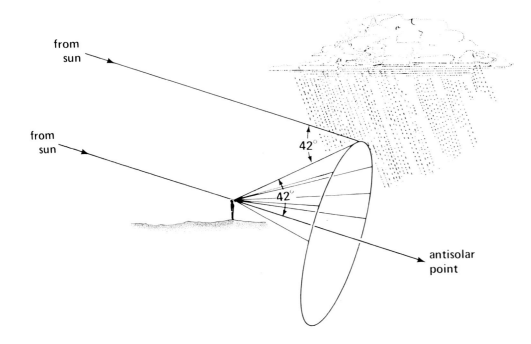

from sun

from sun

42°

42°

antisolar point

Although beams of wavelengths corresponding to all seven colors of the spectrum emerge from each drop, only one of those beams reaches an observer's eye; the beams of the other six wavelengths fall above or below it. The rainbow's seven concentric, semicircular bands of color, with violet at the bottom and red at the top, are created in the eye of the beholder by light coming from raindrops at differing distances from the eye. All of the raindrops transmitting a given wavelength of light will be approximately equidistant and will thus appear to be arranged in a semicircle.

In theory, a rainbow is the visible arc of a more or less perfect circle, the remainder of which is blocked from view by the horizon. (From an airplane a lucky observer may have the good fortune to look down upon a fully circular rainbow with the shadow of the plane visible on a cloud at the rainbow's center.) The rays from the seven bands of the rainbow form seven concentric cones, each having its apex at the observer's eye. The center of all these cones is the line that extends from the sun, through the back of the observer's head to his or her eye, then through the shadow cast by the observer's head, and thence on to a theoretical point below the surface of the earth, the so-called anti-solar point. The angle between that line and the red band of the rainbow is approximately 42°; the angle between the line and the violet band is approximately 40°.

A rainbow can occur only when the sun is less than 42° above the horizon—in other words, before mid-morning or past mid-afternoon. At midday a rainbow will not be visible to an observer on the ground, since the entire bow will be below the horizon. Rainbows are most frequently seen in the late afternoon; they occur then in the eastern part of the sky, with the sun in the west. The smallest arc will be seen when the sun is closest to 42° above the horizon. The fullest arc, with its top at about 42° above the horizon, will occur when the sun has just risen or is about to set. If the conditions necessary for a rainbow persist for a sufficient time during the afternoon, one may observe a rainbow rising in the sky as the sun declines.

When a fainter, secondary bow is visible, it is always approximately eight degrees above the primary one. The primary bow is about two degrees wide. The secondary bow is slightly wider. Sometimes additional partial bows are visible; these are called spurious, supernumerary, or complementary rainbows. A ray of

Tom Bean, *A glory forms around the photographer's shadow, Canyon Rim, Colorado National Monument,* 1994

light may, so to speak, bounce around inside a raindrop as it undergoes a whole series of internal reflections. The rainbows that would theoretically be produced by three or four reflections are almost never visible, for in those cases the rays of light are directed back toward the sun, whose intense light obliterates the faint bows. However, pale segments of the supernumerary bow produced by five reflections may occasionally be glimpsed; they occur below the primary rainbow. Six or more reflections are visible only in single-drop experiments.

In the secondary bow, the red band is always at the bottom and violet at the top, the reverse of the primary bow. The secondary bow is created by light entering the lower part of each raindrop and undergoing two refractions and two reflections. It is less than half as bright as the primary bow, for the intensity of light is reduced by each reflection. The sky between the primary and secondary bows is darker than the surrounding sky. This dark area is called Alexander's band in honor of Alexander of Aphrodisias, who first pointed out the paradox that the area between the red (i.e., strongest) bands of the primary and secondary bows is darkened rather than brightened. In fact, the sky below the primary bow is lightened, for rays emerging from raindrops at angles of less than 42° tend to mix together to form white light.

Sometimes the blurring of the colors is so extreme—especially if a thin cloud filters the light between the sun and the rain, or if the drops are very small, as in fog or in a fine mist—that the entire rainbow appears white. Such a phenomenon is called a "fog bow" or "Ulloa's circle," after Antonio de Ulloa, who described such a bow that he saw in Peru in 1748.

PAUL VUCETICH, *DOUBLE RAINBOW OVER THE NORTH RIM, GRAND CANYON*

GEORGE BELLOWS, *FOG RAINBOW*, 1913

Meteorologia Aristotelis. Eleganti

Iacobi Fabri Stapulensis Paraphrasi explanata.
Comentarioq̃ Ioannis Coclæi Norici declarata
ad fœlices in philosophiæ studiis successus
Calcographiæ iamprimū demandata.

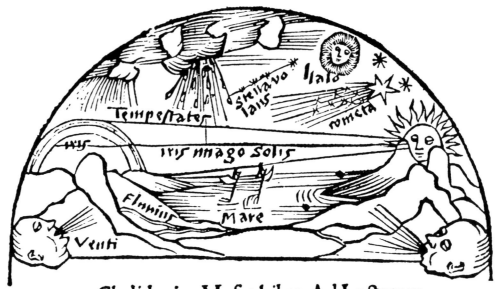

Chelidonius Musophilus. Ad Lectorem.

Est citra decimæ quicquid curuamina sphæræ
    Vna non facie siue tenore manet.
Sedibus ipsa suis mutantur sidera & orbes
    Induit & vultus Luna subinde nouos
Plus etiam phœbes sub fornice.quattuor inter
    Corpora/iuris habet lis & amicicia.
Aeris hinc vastiq̃ maris.telluris & ignis
    Apparent miris plæraq̃ monstra modis
Quæ tibi præsenti Lector cernenda libello
    Producit Cocles eruta mille locis.

TITLE PAGE OF JACQUES LEFEVRE D'ETAPLES'S PARAPHRASE OF ARISTOTLE'S *METEOROLOGICA*, 1512

ONE OF THE EARLIEST ATTEMPTS to give a scientific expla- nation of the rainbow was made by Aristotle (384-322 B.C.) in his book *Meteorologica*. Believing that we see by means of rays projected outward by our eyes, he claimed that our vision is reflected by individual drops of rain. He went on to say that any such tiny "mirror" reflects colors but not shapes. "Each of the reflecting particles is invisibly small, and the continuous magnitude formed by them all is what we see; what appears to us is therefore necessarily a continuous magnitude of a single color." Characteristically, Aristotle (who maintained that three was the dominant operative number in all of nature) distinguished only three colors in the rainbow: red, green, and blue. He postulated that bright light reflected by a dark surface will look red, for reflection diminishes the intensi- ty of white light and thereby creates colors. "In the primary rainbow the outermost band is red, for the vision is reflected most strongly on to the sun from the largest circumference, and the outermost band is the largest." According to Aristotle, the successively diminishing size of the inner bands weakens the intensity of reflection and thus causes the red to change first to green and then, weakest of all, to blue. "There is no further change of color, the complete process consisting, like most oth- ers, of three stages.... The yellow color that appears in the rain- bow is due to the contrast of two others; for red in contrast to green appears light yellow." Because of the reverence with which Aristotle was widely regarded, his theories did much to block the progress of science for nearly two thousand years.

The Roman Stoic philosopher Seneca believed that the rainbow is formed by the reflection of sunlight by a concave cloud, which focuses the rays as a concave mirror would do. During the Middle Ages, writers on the rainbow were divided by their allegiance either to Seneca's cloud theory or to Aristotle's theory of individual raindrops. The writings of other ancient authors on the subjects of optics and meteorology were trans- mitted to northern Europe primarily through the works of Arabic scholars. The most important contribution was made by Alhazen, a great mathematician and physicist who died in Cairo in 1039. His treatise on optics, which was translated into Latin in 1270, included sections dealing with reflection and refraction and with the rainbow.

It was, however, a fourteenth-century German monk known as Dietrich von Freiberg who seems to have been the first to

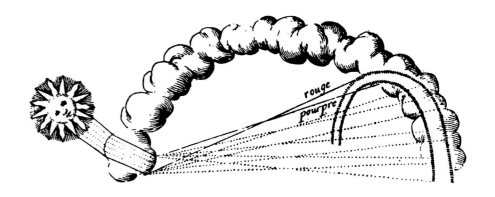

THE FORMATION OF THE RAINBOW, FROM MARIN CUREAU DE LA CHAMBRE'S *NOUVELLES OBSERVATIONS ET CONIECTURES SUR L'IRIS*, 1650

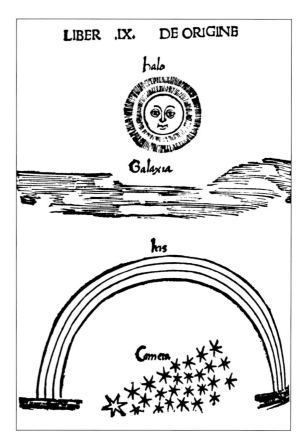

WOODCUT FROM GREGOR REISCH'S *MARGARITA PHILOSOPHICA*, 1503

Assumptio densa Luminis, à medio humentis assumpti : Lumen conspicuum , propter densum, & aliquas tenebras oppositas.

Aer Humidus.

Oculus

Communis aër sed non ita pla

lume assumens; ne , ut referat.

Lux non as transiens.

sumpta, sed

THE FORMATION OF THE RAINBOW, IN WILLIAM GILBERT'S *DE MUNDO NOSTRO SUBLUNARI PHILOSOPHIA NOVA*, 1651

understand that the primary rainbow is caused by two refractions and one reflection within each raindrop, and that the secondary bow is caused by two refractions and two reflections. Dietrich (who is also called Theodoric) arrived at his discovery during the first decade of the fourteenth century, but his manuscripts remained virtually unknown until they were published in 1914.

In the realm of pseudo-science, the rainbow was an important symbol in the lore of alchemy, for it represented the end of the storms of disharmony and darkness. The rainbow was the sign of the advent of the light, under whose reign of peace the base elements could at last be transmuted into gold. The alchemists even believed that just before the physical transformation would take place, the molten metals would give off an efflorescence of colors suggestive of the celestial rainbow.

The year 1611 saw the publication of a treatise on the rainbow that had been written in the 1590s by Marcus Antonius de Dominis, Archbishop of Split, in Croatia. According to Isaac Newton, who was unaware of Dietrich von Freiberg and had evidently not read de Dominis very carefully, the latter was the first to understand that light is refracted twice and reflected once within each drop to produce the primary rainbow. In fact, de Dominis mistakenly believed that the light underwent only one refraction and one reflection.

MICHAEL MAIER, *THE SIXTH KEY: MARRIAGE*. ILLUSTRATION FOR A WORK ON ALCHEMY, 1618

THE FORMATION OF THE RAINBOW, FROM A MANUSCRIPT COPY OF DIETRICH VON FREIBERG'S TREATISE, EARLY 14TH CENTURY

39

40

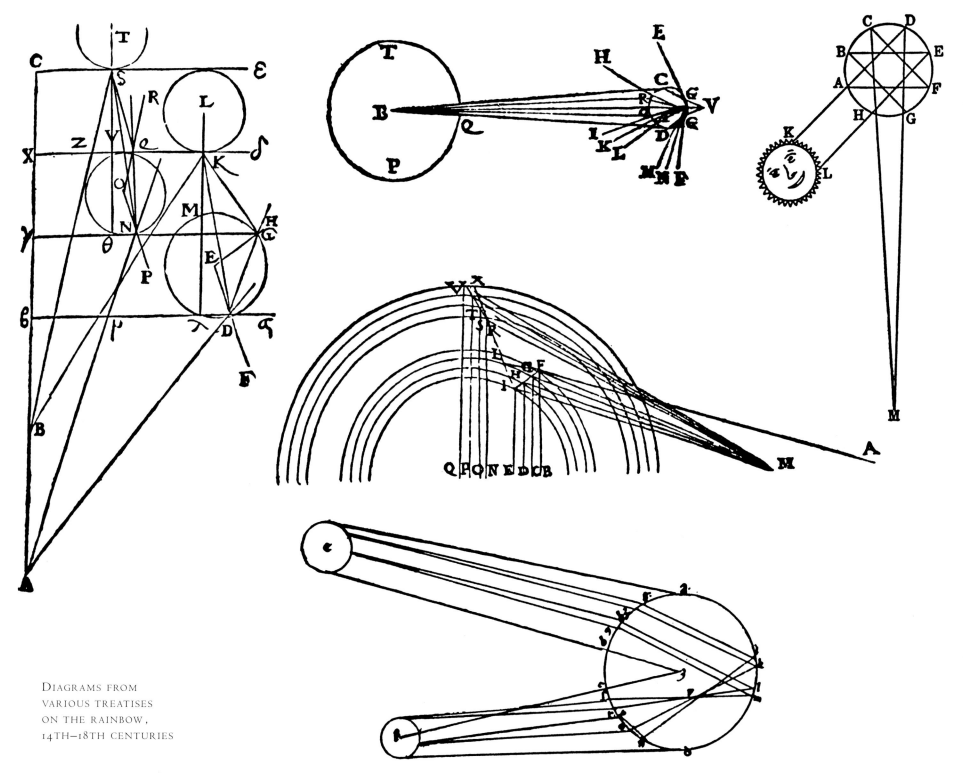

ALBRECHT DÜRER, *MELANCHOLIA I,* 1514

41

From Descartes's *Discourse on Method*:

Firstly, having considered that the rainbow can appear not only in the sky but also in the air near us, whenever there are drops of water illuminated by the sun, as can sometimes be seen in fountains, I easily understood that the rainbow is created only by the way in which the rays of light act upon these drops and are then presented to our eyes. Furthermore, knowing that raindrops are round, ... and seeing that it does not matter whether these drops are large or small in order to produce a rainbow, I decided to make a large drop, in order to be able to examine it well. To this end, I filled with water a large, perfectly round, and very transparent glass flask. I then found that when I put this sphere at the place BCD (with the sun coming from the part of the sky marked AFZ and my eye at the point E), the point D appeared to me completely red and incomparably brighter than the rest of the flask. Whether I approached it or withdrew from it, and whether I moved it to the right or the left, ... provided that the line DE always made an angle of 42 degrees with the line EM—the latter being imagined to extend from the center of the sun [and through the back of my head] to the center of my eye—this point D always remained equally red. But as soon as I made the angle DEM either smaller or larger, this redness disappeared; and, if I moved the sphere a little less, the redness did not disappear all at once but first divided itself into two less brilliant parts, in which could be seen yellow, blue, and other colors.

Then, looking also toward the part of the sphere marked K, I perceived that, when the angle KEM was about 52 degrees, K also appeared red, but not so bright as D. When I made the angle a bit larger, other colors appeared more weakly; but when I made it either a little smaller or much larger, no colors appeared. From this I knew manifestly that if all the air towards M were filled with such spheres, or with drops of water, there would appear a very bright red point in each of these drops from which the lines drawn toward the eye E make an angle of about 42 degrees with EM, as I suppose those marked R would do. These points, being regarded all together, ... must appear as a continuous circle of red; and there must be, in just the same manner, points in those drops marked S and T, from which the lines drawn to the eye E form angles a bit more acute than EM, which compose circles of weaker colors. It is of these that the primary and principal rainbow consists.

Then, once again, the angle MEX being 52 degrees, there must appear a red circle in the drops marked X, and other circles of weaker colors in the drops marked Y, and it is of these that the secondary and lesser rainbow consists. Finally, in all the other drops marked V, no colors can appear.

Examining, thereafter, more particularly what made the point D red on the sphere BCD, I found that the rays of the sun, coming from A toward B, curved upon entering the water at point B and continued on toward C, where they were reflected toward D, and there, curving once again upon leaving the water, extended toward the eye E. I could be certain of this because, as soon as I placed an opaque or dark object anywhere along the lines AB, BC, CD, or DE, the red color disappeared. And if I covered the entire sphere except for the points B and D, and put the dark objects anywhere else, as long as nothing interrupted the lines ABCDE, the red did not cease to appear.

Then, seeking also to discover why red appeared at K, I found that this was caused by the rays coming from F toward G, where they bent toward H, and at H they were reflected toward I, and at I were reflected once again toward K, then finally curving at point K and extending toward E. Therefore, the primary rainbow is caused by the rays that come to the eye after two refractions and one reflection, and the secondary rainbow is caused by other rays which arrive at the eye only after two refractions and two reflections, which prevents the rays from being as strong as those of the primary rainbow.

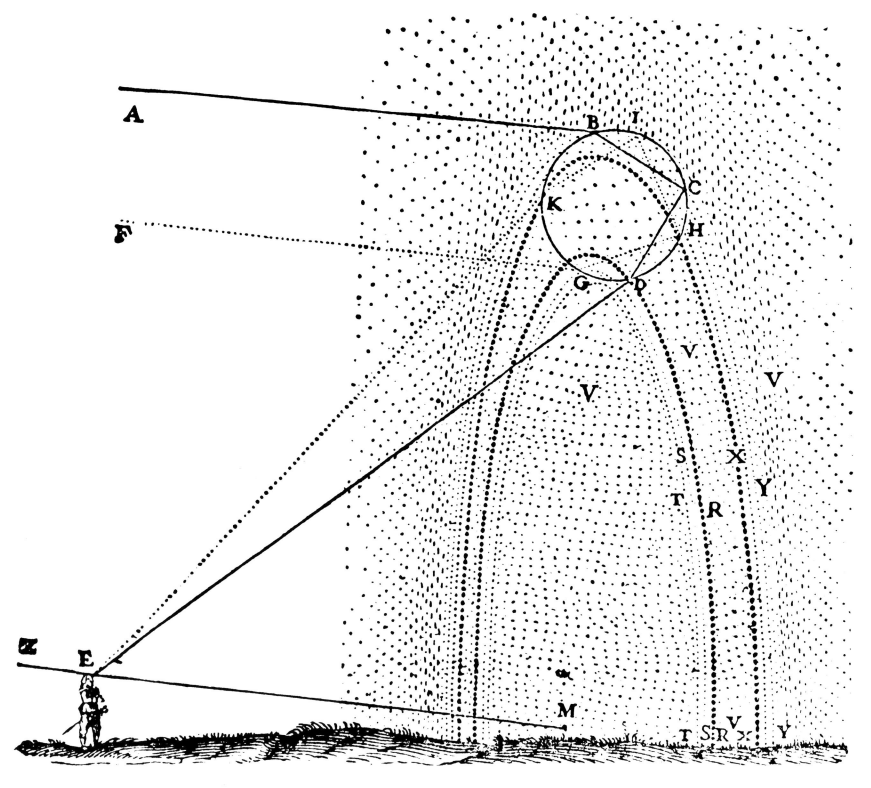

PETER FLÖTNER, *SEVERAL WAYS OF LIGHT REFRACTION*, 1535

The correct explanation was first published by the seventeenth-century French philosopher and mathematician René Descartes, who is best known for his deduction "I think; therefore, I am." Descartes wrote an important treatise on the rainbow, which he included in *Les Météores* (as in meteorology), an appendix to his great *Discours de la Méthode pour Bien Conduire Sa Raison et Chercher la Verité dans les Sciences* (Discourse on the Method of Guiding the Reason and Searching for Truth in the Sciences), published in 1637. He began by stating, "The rainbow is a marvel of nature so remarkable, and its cause has been curiously researched by the best minds of all times, and so little understood, that I do not know how I could choose a subject more suited to show how, by means of the method I use, one can arrive at knowledge that was not possessed by those whose writings have come down to us." He then went on to give an essentially accurate analysis of the basic mechanics of the rainbow.

Descartes's calculations were further developed by two Dutchmen, Christian Huygens and Benedictus de Spinoza. The latter, one of the greatest philosophers in the history of Western thought, was a lens grinder by profession; consequently, he was expert in questions of refraction and reflection. The method by which he arrived at a radius of 40° 57' for the primary bow and 54° 25' for the secondary was published in 1687, ten years after his death, as *The Algebraic Calculation of the Rainbow*.

It was Isaac Newton who, in 1666, first understood that white light is composed of a continuous spectrum of seven colors and an infinite number of gradations between them. It was also Newton who first propounded the wave theory of light, refuting the erroneous particle theory espoused by his predecessors. Violet is at the short-wavelength (high-frequency) end of the visible spectrum, beyond which lies ultraviolet; red is at the long-wavelength (low-frequency) end, beyond which lies infrared.

The most important work in the entire history of the scientific understanding of the rainbow is Newton's book *Opticks: or, A Treatise of the Reflexions, Refractions, Inflexions and Colours of Light*, first published in 1704 and subsequently issued in numerous revised and enlarged editions. Only about ten pages are devoted specifically to the rainbow (Book One, Part II, Proposition IX, Problem IV: By the discovered Properties of Light to explain the Colours of the Rainbow), but Newton's calculations and conclusions were more precise than those of Descartes.

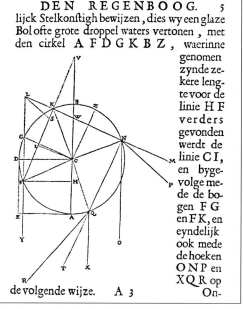

PAGES FROM SPINOZA'S TREATISE ON THE RAINBOW, 1687, AND NEWTON'S *OPTICKS*, 1704

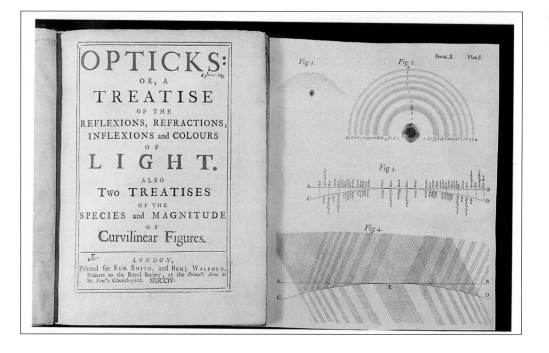

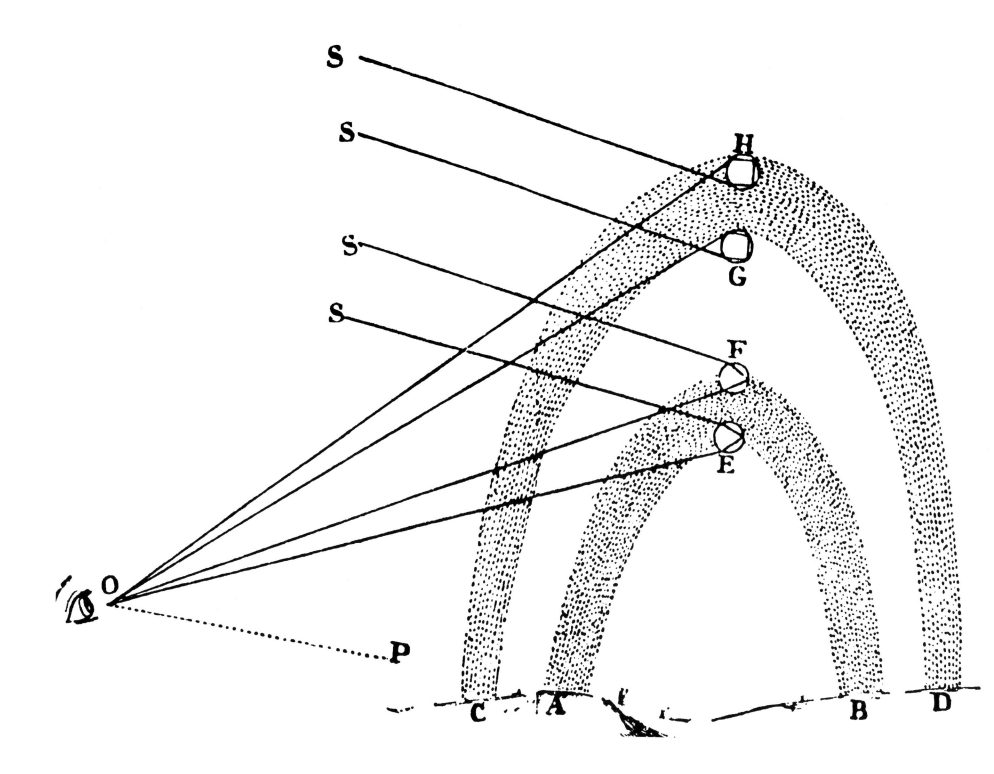

From Newton's *Opticks*:

Suppose now that O is the Spectator's Eye, and OP a Line drawn parallel to the Sun's Rays and let POE, POF, POG, POH, be Angles of 40 Degrees 17 Minutes, 42 Degr. 2 Min., 50 Degr. 57 Min., and 54 Degr. 7 Min. respectively, and these Angles turned about their common Side OP, shall with their other Sides OE, OF, OG, OH, describe the Verges of two Rainbows AF, BE and CHDG. For if E, F, G, H, be drops placed any where in the conical Superficies described by OE, OF, OG, OH, and be illuminated by the Sun's Rays SE, SF, SG, SH; the Angle SEO being equal to the Angle POE, or 40 Degr. 17 Min. shall be the greatest Angle in which the most refrangible Rays can after one Reflexion be refracted to the Eye, and therefore all the Drops in the Line OE shall send the most refrangible Rays most copiously to the Eye, and thereby strike the Senses with the deepest violet Colour in that Region.

And in like manner the Angle SFO being equal to the Angle POF, or 42 Degr. 2 Min., shall be the greatest in which the least refrangible Rays after one Reflexion can emerge out of the Drops, and therefore those Rays shall come most copiously to the Eye from the Drops in the Line OF, and strike the Senses with the deepest red Colour in that Region. And by the same Argument, the Rays which have intermediate Degrees of Refrangibility shall come most copiously from Drops between E and F, and strike the Senses with the intermediate Colours, in the order which their Degrees of Refrangibility require, that is in the Progress from E to F, or from the inside of the Bow to the outside in this order: violet, indigo, blue, green, yellow, orange, red. But the violet, by the mixture of the white Light of the Clouds, will appear faint and incline to purple.

Again, the Angle SGO being equal to the Angle POG, or 50 Degr. 57 Min., shall be the least Angle in which the least refrangible Rays can after two Reflexions emerge out of the Drops, and therefore the least refrangible Rays shall come most copiously to the Eye from the Drops in the Line OG, and strike the Senses with the deepest red in that Region. And the Angle SHO being equal to the Angle POH, or 54 Degr. 7 Min., shall be the least Angle, in which the most refrangible Rays after two Reflexions can emerge out of the Drops; and therefore those Rays shall come most copiously to the Eye from the Drops in the Line OH, and strike the Senses with the deepest violet in that Region. And by the same Argument, the Drops in the Regions between G and H shall strike the Sense with the intermediate Colours in the order which their Degrees of Refrangibility require, that is, in the Progress from G to H, or from the inside of the Bow to the outside in this order: red, orange, yellow, green, blue, indigo, violet. And since these four Lines OE, OF, OG, OH, may be situated any where in the above-mention'd conical Superficies; what is said of the Drops and Colours in these Lines is to be understood of the Drops and Colours every where in those Superficies.

Do not all charms fly

At the mere touch of cold philosophy?

There was an awful rainbow once in heaven:

We know her woof, her texture; she is given

In the dull catalogue of common things.

Philosophy will clip an angel's wings,

Conquer all mysteries by rule and line,

Empty the haunted air, and gnomed mine —

Unweave a rainbow.

—John Keats, *Lamia,* 1820

Keats once proposed a toast, "Newton's health, and confusion to mathematics," on the grounds that Newton had "destroyed the poetry of the rainbow by reducing it to a prism."

JOHN CONSTABLE, *LONDON FROM HAMPSTEAD, WITH A DOUBLE RAINBOW*, 1831

We shall find the cube of the rainbow,
Of that there is no doubt;
But the arc of a lover's conjecture
Eludes the finding out.

—Emily Dickinson (1830–1866)

The rainbow held an unexpected fascination for two of the most austerely Calvinistic of American Puritan ministers, Cotton Mather (1663-1728) and Jonathan Edwards (1703-1758). In 1712 Mather—who became a corresponding member of the most distinguished British scientific body, the Royal Society—published a 64-page book entitled *Thoughts for the Day of Rain,* the two sections of which are called "The Gospel of the Rainbow" and "The Saviour with His Rainbow." Mather had boundless admiration for Newton, whom he called "The Perpetual Dictator of the Learned World in the Principles of Natural Philosophy; and the most Sagacious Reasoner upon the Laws of Nature that has yet Shone among Mankind." Although Mather was well acquainted with the works of de Dominis and Descartes, as well as with what he referred to as Newton's "Incomparable Treatise of Opticks," he was most interested in the power of the rainbow to inspire pious thoughts. He advocated

> such an Useful Contemplation of the Rainbow as may render it a noble Instrument of Piety, in our whole Conversation. I shall not Enquire into the old opinion that the Spots of Earth, where the Feet of the Rainbow strike, are made singularly Fruitful by the Influences of it. But I am sure, the Rainbow striking on our Minds may produce many Sweet Fruits of Pious Devotion.

Mather found, for instance, an "Instruction of Humility" in the observation "The higher the Sun, the lesser the Rainbow." From this he extrapolated, "The Higher a Glorious CHRIST is with us, and in us, and the more He does for us, the Smaller must we be in our own Eyes."

Edwards, who is remembered for his terrifying sermon entitled "Sinners in the Hands of an Angry God," had a precocious interest in natural history and, when he was about eleven years old, wrote an essay on the Newtonian theory of the rainbow. He declared himself "fully satisfied of Sir Isaac Newton's different reflexibility and refrangibility of the rays of light," and he went on to state:

> It cannot be the cloud from whence this reflection is made, as was once thought, for we almost alwaies see the ends of rainbows come down even in amongst the trees below the hills, and to the very ground where we know there is no part of the cloud there, but what descends in drops of rain. I can convince any man by ocular demonstration in two minutes on a fair day that the reflection is from drops by only taking a little water in my mouth and standing between the sun and something that looks a little darkish and spirting of it into

the air so as to disperse all into fine drops, and there will appear as compleat and plain a rainbow with all the colours as ever was seen in the heavens.

Not everyone was enthusiastic about Newton. The great German poet and polymath Johann Wolfgang von Goethe condemned the Newtonian optics as "cunning lawyers' tricks and sophistical distortions of nature." Goethe's own contributions to the subject were his books *Beiträge zur Optik* (Studies Regarding Optics), published in 1791, and *Der Farbenlehre* (Theory of Color), 1805-10. Late in life he wrote, "As for what I have done as a poet I take no pride whatever in that.... But in my century I am the only person who knows the truth in the difficult science of colors, ... of that I say I am not a little proud."

During the nineteenth and twentieth centuries, scientific investigations of the rainbow have been concerned mainly with refinements of Newton's theories in the abstruse regions of optics, physics, and mathematics. However, one breakthrough quite comprehensible to the layperson was made in 1966 by Robert Greenler, professor of physics at the University of Wisconsin in Milwaukee. Greenler assumed that in every rainbow there must be, adjacent to the red band, an infrared band, which, though invisible to the naked eye, could perhaps be recorded on special black-and-white film sensitive to infrared rays. (Most ultraviolet rays, much weaker than the infrared, are absorbed by the atmosphere and by the raindrops; therefore, even if there were an ultraviolet band, it would be far too faint to photograph.)

"The problem in taking a picture with the infrared film," Greenler wrote, "is that it is sensitive not only to the infrared but to all of the visible, and it is extremely sensitive to blue light. An unfiltered exposure would result in a black-and-white negative, but there would be no way to tell what part of the spectrum produced what part of the exposure." Consequently, he used an opaque black filter that would allow only the infrared rays to reach the film. In 1966 he successfully recorded the infrared band of a rainbow that he produced artificially with a perforated garden hose. Four years later he photographed the infrared bands of a natural rainbow and of its secondary bow as well. "When the film was developed," wrote Greenler, "I saw for the first time an infrared rainbow that had hung in the sky, undetected, since before the presence of people on this planet."

ROBERT GREENLER, *INFRARED PHOTOGRAPHS OF A RAINBOW*, 1966 AND 1970

Eighteenth and Nineteenth Centuries

On yonder lofty mountain
A thousand times I stand,
And on my staff reclining,
Look down on smiling land.

My grazing flocks then I follow,
My dog protecting them well; 73

I find myself in the valley,
But how, I scarcely can tell.

The whole of the meadow is cover'd
With flowers of beauty rare;
I pluck them, but pluck them unknowing
To whom the offering to bear.

In rain and storm and tempest,
I tarry beneath the tree,
But closed remaineth yon portal;
'Tis all but a vision to me.

High over yonder dwelling,
There rises a rainbow gay;
But she from home hath departed,
And wander'd far, far away.

—Johnann Wolfgang von Goethe, *The Shepherd's Lament*, 1802

Caspar David Friedrich, *Mountain Landscape with Rainbow*, 1810

Caspar Wolf,
Snow Bridge and Rainbow on the Gadmental, 1778

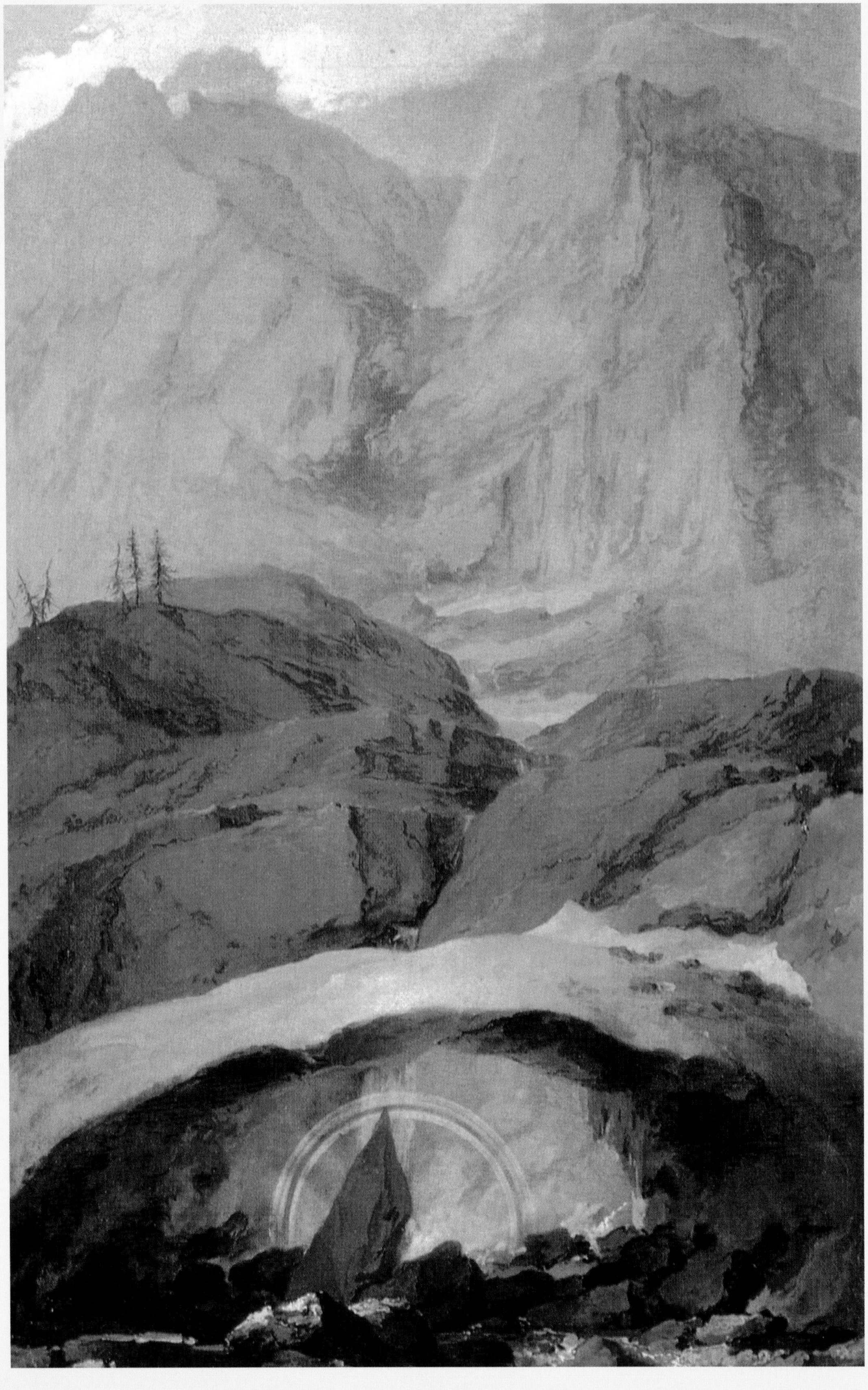

Ye Ice-falls! ye that from the mountain's brow

Adown enormous ravines slope amain —

Torrents, methinks, that heard a mighty voice,

And stopped at once amid their maddest plunge!

Motionless torrents! silent cataracts!

Who made you glorious as the gates of Heaven

Beneath the keen full moon? Who bade the sun

Clothe you with rainbows? Who, with living flowers

Of loveliest blue, spread garlands at your feet?

—Samuel Taylor Coleridge (1772–1834), *Hymn Before Sunrise, Vale of Chamouni*

The Rainbow comes and goes,

And lovely is the Rose,

The Moon doth with delight

Look round her when the heavens are bare,

Waters on a starry night

Are beautiful and fair;

The sunshine is a glorious birth;

But yet I know, where'er I go,

That there hath past away a glory from the earth.

—William Wordsworth, *Ode: Intimations of Immortality, 1803–07*

JOSEPH ANTON KOCH, *LANDSCAPE AFTER A THUNDERSTORM*, C. 1830   57

KARL FRIEDRICH SCHINKEL, *GOTHIC CATHEDRAL WITH IMPERIAL PALACE*, 1815

JOHN CONSTABLE, *SALISBURY CATHEDRAL FROM THE MEADOWS*, 1831

JOHN CONSTABLE, *THE GROVE, OR ADMIRAL'S HOUSE, HAMPSTEAD*, 1821–22   JOHN CONSTABLE, *LANDSCAPE AND DOUBLE RAINBOW*, 1812

Now overhead a rainbow, bursting through

The scattering clouds, shone, spanning the dark sea,

Resting its bright base on the quivering blue,

And all within its arch appeared to be

Clearer than that without, and its wide hue

Waxed broad and waving, like a banner free,

Then changed like to a bow that's bent, and then

Forsook the dim eyes of those shipwrecked men.

It changed of course — a heavenly chameleon,

The airy child of vapour and the sun,

Brought forth in purple, cradled in vermilion,

Baptized in molten gold and swathed in dun,

Glittering like crescents o'er a Turk's pavilion

And blending every colour into one ...

Our shipwrecked seamen thought it a good omen;

It is as well to think so now and then.

'Twas an old custom of the Greek and Roman,

And may become of great advantage when

Folks are discouraged; and most surely no men

Had greater need to nerve themselves again

Than those, and so this rainbow looked like hope,

Quite a celestial kaleidoscope.

—George Noel Gordon, Lord Byron, *Don Juan, Canto II*, 1819

Francis Danby (1793–1861), *The Shipwreck*, n.d.

Triumphal arch that fills the sky
When storms prepare to part,
I ask not proud Philosophy
To teach me what thou art —

Still seem, as to my childhood's sight
A midway station given
For happy spirits to alight
Betwixt the earth and heaven.

Methinks, thy jubilee to keep,
The first-made anthem rang
On earth deliver'd from the deep,
And the first poet sang.

Nor ever shall the Muse's eye
Unraptured greet thy beam;
Theme of primeval prophecy
Be still the prophet's theme!

—Thomas Campbell, *To the Rainbow*, 1819

J.M.W. TURNER (1775–1851), *RIVER SCENE WITH RAINBOW*, N.D.

J.M.W. Turner, *Rainbow: Osterspey and Feltzen on the Rhine*, c. 1820

Compressed as in a tunnel: from the lake
Bodies of foam took flight, and everything
Was wrought into commotion, high and low —
A roaring wind, mist, and bewildered showers.
Ten thousand thousand waves, mountains and crags.
And darkness, and the sun's tumultuous light.
Green leaves were rent in handfuls from the trees …
The horse and rider staggered in the blast …
Meanwhile, by what strange chance I cannot tell,
What combination of the wind and clouds,
A large unmutilated rainbow stood
Immovable in heaven.

—William Wordsworth (1770–1850), *Prelude*

. . .[S]uddenly
The rain and the wind ceased, and the sky
Received at once the full fruition
Of the moon's consummate apparition.
The black cloud-barricade was riven,
Ruined beneath her feet, and driven
Deep in the West; while, bare and breathless,
North and South and East lay ready
For a glorious thing that, dauntless, deathless,
Sprang across them and stood steady.
'Twas a moon-rainbow, vast and perfect,
From heaven to heaven extending, perfect
As the mother-moon's self, full in face.
It rose, distinctly at the base
With its seven proper colors chorded,
Which still, in the rising, were compressed,

Until at last they coalesced,
And supreme the spectral creature lorded
In a triumph of whitest white,—
Above which intervened the night.
But above night too, like only the next,
The second of a wondrous sequence,
Reaching in rare and rarer frequence,
Till the heaven of heavens were circumflexed,
Another rainbow rose, a mightier,
Fainter, flushier and flightier,—
Rapture dying along its verge.
Oh, whose foot shall I see emerge,
Whose, from the straining topmost dark,
On to the keystone of that arc?

—Robert Browning, *Christmas Eve,* 1850

Joseph Wright of Derby, *Landscape with a Rainbow*, 1794

69

Or, if a sudden silver shower

Has drench'd the molten sunset hour,

And with weeping cloud is spread

All the welkin overhead,

Save where the unvexèd west

Lies divinely still, at rest,

Where liquid heaven sapphire-pale

Does into amber splendours fail,

And fretted clouds with burnish'd rim,

Phoebus' loosen'd tresses, swim;

While the sun streams forth amain

On the tumblings of the rain,

When his mellow smile he sees

Caught on the dank-ytressèd trees,

When the rainbow arching high

Looks from the zenith round the sky,

Lit with exquisite tints seven

Caught from angels' wings in heaven,

Double, and higher than his wont,

The wrought rim of heaven's font, —

Then may I upwards gaze and see

The deepening intensity

Of the air-blended diadem,

All a sevenfold-single gem,

Each hue so rarely wrought that where

It melts, new lights arise as fair,

Sapphire, jacinth, chrysolite,

The rim with ruby fringes dight,

Ending in sweet uncertainty

'Twixt real hue and phantasy.

Then while the rain-born arc glows higher

Westward on his sinking sire;

While the upgazing country seems

Touch'd from heaven in sweet dreams;

While a subtle spirit and rare

Breathes in the mysterious air;

While sheeny tears and sunlit earth

Mix o'er the not unmovèd earth, —

Then would I fling me up to sip

Sweetness from the hour, and dip

Deeply in the archèd lustres,

And look abroad on sunny clusters

Of wringing tree-tops, chalky lanes,

Wheatfields tumbled with the rains,

Streaks of shadow, thistled leas,

Whence spring the jewell'd harmonies

That meet in mid-air; and be so

Melted in the dizzy bow

That I may drink that ecstacy

Which to pure souls alone may be. . . .

—Gerald Manley Hopkins, *II Mistico*, 1880s

I bind the sun's throne with a burning zone,

And the moon's with a girdle of pearl;

The volcanoes are dim, and the stars reel and swim

When the whirlwinds my banner unfurl.

From cape to cape, with a bridge-like shape,

Over a torrent sea,

Sunbeam-proof, I hang like a roof,

The mountains its columns be.

The triumphal arch through which I march,

With hurricane, fire, and snow,

When the powers of the air are chained to my chair,

Is the million-coloured bow;

The sphere-fire above its soft colours wove,

While the moist earth was laughing below.

—Percy Bysshe Shelley (1792–1822)

FREDERIC EDWIN CHURCH, *RAINY SEASON IN THE TROPICS*, 1866

FREDERIC EDWIN CHURCH, *THUNDERSTORM IN THE ALPS*, 1868

Nor even yet

The melting rainbow's vernal-tinctur'd hues

To me have shone so pleasing, as when first

The hand of science pointed out the path

In which the sun-beams gleaming from the west

Fall on the watry cloud, whose darksome veil

Involves the orient, and that trickling show'r

Piercing thro' every crystalline convex

Of clust'ring dew-drops to their flight oppos'd,

Recoil at length where concave all behind

Th' internal surface of each glassy orb

Repells their forward passage into air;

That thence direct they seek the radiant goal

From which their course began; and, as they strike

In diff'rent lines the gazer's obvious eye,

Assume a diff'rent lustre, thro' the brede

Of colours changing from the splendid rose

To the pale violet's dejected hue.

—Mark Akenside, *The Pleasures of Imagination*, 1744

ALVAN FISHER, *THE GREAT HORSESHOE FALLS, NIAGARA*, 1820

John Trumbull, *Niagara Falls from below the Great Cascade on the British Side*, c. 1807

*A*nd God smiled again,
And the rainbow appeared,
And curled itself around his shoulder.

—James Weldon Johnson, *God's Trombones; The Creation*, 1927

ROBERT SCOTT DUNCANSON, *LANDSCAPE WITH RAINBOW*, 1859

GEORGE INNESS, *THE RAINBOW*, C. 1878–79

GEORGE INNESS, *A PASSING SHOWER*, 1860

Look upon the rainbow, and praise Him that made it.

—Ecclesiastes 43:2

ALBERT BIERSTADT, *RAINBOW OVER JENNY LAKE*, C. 1870

Tom Bean, *Rainbow at Yaki Point on South Rim, Grand Canyon National Park, Arizona*, 1990

Edward Potthast, *The Grand Canyon*, 1910

God loves an idle rainbow/No less than laboring seas.

—Ralph Hodgson (1871–1962), *A Wood Song*

JASPER FRANCIS CROPSEY,
THE NARROWS FROM STATEN ISLAND, 1866–68

I understand how scarlet can differ from crimson because I know that the smell of an orange is not the smell of a grapefruit. I can also conceive that colors have shades and guess what shades are. In smell and taste there are varieties not broad enough to be fundamental; so I call them shades... The force of association drives me to say that white is exalted and pure, green is exuberant, red suggests love or shame or strength. Without the color or its equivalent, life to me would be dark, barren, a vast blackness.

Thus through an inner law of completeness my thoughts are not permitted to remain colorless. It strains my mind to separate color and sound from objects. Since my education began I have always had things described to me with their colors and sounds, by one with keen senses and a fine feeling for the significant. Therefore, I habitually think of things as colored and resonant. Habit accounts for part. The soul sense accounts for another part. The brain with its five-sensed construction asserts its right and accounts for the rest. Inclusive of all, the unity of the world demands that color be kept in it whether I have cognizance of it or not. Rather than be shut out, I take part in it by discussing it, happy in the happiness of those near to me who gaze at the lovely hues of the sunset or the rainbow.

—Helen Keller, *The Story of My Life*, 1902

JOHN EVERETT MILLAIS, *THE BLIND GIRL*, 1856

Ford Madox Brown, *Walton-on-the-Naze*, 1859–60

CAMILLE PISSARRO, *LA PLAINE D'EPLUCHES (RAINBOW)*, 1877

Till, in the western sky, the downward sun
Looks out effulgent from amid the flush
Of broken clouds, gay-shifting to his beam.
The rapid radiance instantaneous strikes
The illumined mountain, through the forest streams
Shakes on the floods, and in a yellow mist,
Far smoking o'er the interminable plain,
In twinkling myriads lights the dewy gems.
Moist, bright, and green, the landscape laughs aound.
Full swell the woods; their every music wakes,

Mixed in wild concert, with the warbling brooks
Increased, the distant bleatings of the hills,
The hollow lows responsive from the vales,
Whence, blending all, the sweetened zephyr springs.
Meantime, refracted from yon eastern cloud,
Bestriding earth, the grand ethereal bow
Shoots up immense; and every hue unfolds,
In fair proportion running from the red
To where the violet fades into the sky.

—James Thomson, *The Seasons: Spring,* 1728

Jean-François Millet, *Springtime*, 1868

CLAUDE MONET, *THE JETTY OF LE HAVRE*, 1868

GEORGES SEURAT, *THE RAINBOW*, 1883

To gild refined gold, to paint the lily,

To throw a perfume on the violet,

To smooth the ice, or add another hue

Unto the rainbow, or with taper-light

To seek the beauteous eye of heaven to garnish,

Is wasteful and ridiculous excess.

—William Shakespeare, *King John*, 1596–97

WILLIAM BLAIR BRUCE, *THE RAINBOW*, C. 1888

You may grind their souls in the self-same mill,

You may bind them, heart and brow;

But the poet will follow the rainbow still,

And his brother will follow the plow.

—John Boyle O'Reilly (1844–1890), *The Rainbow's Treasure*

WILLEM ROELOFS, *THE RAINBOW*, C. 1874-75

We will sing to you all the day:

Mariner, mariner, furl your sails,

For here are blissful downs and dales,

And merrily, merrily carol the gales,

And the spangle dances in bight and bay,

And the rainbow forms and flies on the land

Over the islands free;

And the rainbow lives in the curve of the sand;

Hither, come hither and see;

And the rainbow hangs on the poising wave,

And sweet is the colour of cove and cave,

And sweet shall your welcome be.

—Alfred, Lord Tennyson, *The Sea-Fairies,* 1830

PAUL SIGNAC, *ENTRANCE TO THE PORT OF HONFLEUR,* 1899

As when the rainbow, opposite the sun
A thousand intermingled colors throws.
With saffron wings then dewy Iris flies
Through heaven's expanse, a thousand varied dyes
Extracting from the sun, opposed in place.

—Virgil (70–19 B.C.)

ARKHIP KUINDZHI (1842?–1910), *THE RAINBOW*, N.D.

The Mythology of Rainbows

ONE OF THE MOST widely held traditional beliefs concerning the rainbow is that it is a bridge between heaven and earth, between the divine realm and the human world. The ancient Greeks and Romans associated the rainbow with Iris, one of the chief messengers of the gods. She was especially devoted to the great goddess Juno, who often sent her on errands, some of them rather malicious.

In some passages of the classics Iris seems to be the personification of the rainbow; in others she seems to use the rainbow as a sort of portable road along which to travel. That this was never clearly defined is evident in the following passage from Ovid's *Metamorphoses*, which has it both ways:

> So Iris, in her thousand-colored mantle,
> A rainbow through the sky, sought out the palace
> Under the cloud, the royal home of Sleep....
> 　　　　　The Maiden Goddess
> Entered, using her hands to part the dreams,
> To clear her way, and the shining of her garments
> Brightened the holy home, and the god saw her ...
> And, her instruction given, Iris left him,
> For all too soon the magic spell of slumber
> Was stealing though her limbs, and she soared upward
> Along the rainbow arch she had descended.
> 　　　(Translation by Rolfe Humphries.)

In the *Theogony*, which recounts the history of the Greek gods, the poet Hesiod relates that Iris was the daughter of Thaumas and Elektra and the sister of the Harpies. Thaumas was a sea god, whose name has given rise to the word "thaumaturgy," meaning "miracle-working." (An arcane name for the rainbow in medieval Latin was Thaumantias.) His wife, Elektra (whose name means "bright one"), daughter of Ocean and sister of the river Styx, is not to be confused with the notorious daughter of Agamemnon and Clytemnestra.

In his *Dictionary of the English Language*, published in 1743, Samuel Johnson regarded the words "rainbow" and "iris" as synonymous. Although that usage of "iris" is no longer current in English, the Spanish term for "rainbow" remains "arco iris." In Italian the word "iride" is sometimes used instead of the more common "arcobaleno," which means "lightning bow," in the sense of a bow that would be used by an archer to shoot bolts of lightning.

Most often the rainbow has been thought of as the bridge

ROMAN FRESCO, *JUPITER AND A RAINBOW*, MID-1ST CENTURY A.D.

And Los beheld the mild Emanation Jerusalem eastward bending
Her revolutions toward the Starry Wheels in maternal anguish
Like a pale cloud arising from the arms of Beulahs Daughters:
In Entuthon Benythons deep Vales beneath Golgonooza.

WILLIAM BLAKE, *LOS'S VISION OF BEULAH*, FROM *JERUSALEM*, 1808–1818

105

JOHN FLAXMAN, *IRIS WITH VENUS AND MARS*, 1793

In the fifth book book of Homer's *Iliad*, Venus is wounded in the hand as she attempts to rescue her son Æneas. "To aid her, swift the winged Iris flew," and conducted her to the seat of Mars. Venus then borrowed her brother's chariot, in which Iris, "the goddess of the painted bow," drove her to Juno's throne to ask for revenge.

In Book V of Virgil's *Æneid* the hero's ships are blown off course to Sicily, the location of his father's tomb. Æneas then decides to hold funeral games in his father's honor. When he makes sacrifices before the tomb, an immense but friendly serpent emerges to taste the offerings.

"More various colors thro' his body run,/Than Iris when her bow imbibes the sun." Juno, angry that the games were not dedicated to her, dispatches Iris down her multicolored arch to incite the Trojan women accompanying Æneas to burn their ships. Dreading the hardships of a resumed voyage, the women are susceptible to the spell of Iris, "great in mischief," when she assumes the form of an old woman. But as soon as the imposture is exposed, Iris soars into the heavens upon an immense rainbow. Overwhelmed by this sight, the frenzied women set fire to the ships, though only four are destroyed before Jove extinguishes the flames with rain.

TOM BEAN, *Rainbow at Wukoki, Prehistoric Indian Ruin at Wupatki National Monument, near Flagstaff, Arizona*, 1990

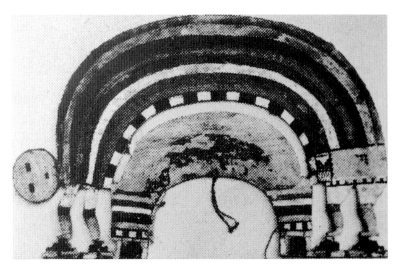

ZUNI RAINBOW HEADDRESS

over which souls travel to heaven, or over which saints travel from heaven to earth. However, Welsh folklore has it that the rainbow is the road by which the Man in the Moon—who was guilty of having worked on Sunday—was forced to travel when he went from earth to his penitential lunar dwelling.

According to the Norse myths recounted in the *Prose Edda* by the Icelandic writer Snorri Sturluson (1179-1241), the rainbow is the bridge to Valhalla. The god Heimdall stands guard over this bridge, known as Bifrost, which flames in three colors—gold, red, and blue. If the frost giants were to try to cross it, they would melt. But it was said that at the Ragnarök, the Twilight of the Gods, the bridge will cool and the giants will storm Valhalla, breaking the bridge into bits.

A Japanese myth relates how, when the world was unformed chaos, the god Izanagi and the goddess Izanami were ordered by their superiors to stand on a rainbow, the "floating bridge of heaven," and from there to stir the ocean with a jeweled spear. The brine that dripped from the spear when it was withdrawn created the first land mass, the island of Onogoro. Izanagi and Izanami then walked down the rainbow bridge called Niji (the Japanese word for "rainbow") to the island, which was eventually divided into the eight islands of Japan and populated with their descendants.

Many of the tribes of the American Southwest look upon the rainbow as the bridge of the gods. The Navajos also believe that the gods chose the rainbow as their conveyance, a sort of flying carpet, because it moves so rapidly. Moreover, the Rainbow Guardian is an essential element of Navaho sandpaintings, which are used in healing ceremonies. Running along three sides of the square designs, the rainbow protects the deities within its embrace.

In Hopi and other Pueblo Indian myths, the Cloud People and the kachinas travel on the rainbow. The ladder down to the subterranean Pueblo ceremonial rooms known as kivas is said to represent the rainbow.

Native American peoples developed an extraordinarily wide variety of legends to account for the rainbow. The Shasta Indians of northwestern California, for instance, believed that Sun used the rainbow colors to paint itself when it came down to the earth as a shaman. The Mojaves of Arizona said the rainbow was a charm the Creator needed to stop a rain storm; a very bad storm required all of the colors. The Hidatsa, a Sioux people of the

BALTHUS, *The Méditerranée's Cat*, 1949

After the Dark Thunder People had fought with the Winter Thunder group until there was great destruction on both sides, the Holy People met to see what could be done. To this meeting at the home of Dark Thunder came Changing Woman, Child-of-the-Water, Bat, Big Fly, and Black God.

At this time two meetings were held simultaneously, one at the home of Dark Thunder and one at the home of Winter Thunder; the same deities were present at both. Although Changing Woman begged the great gods to carry the offer of peace to Winter Thunder, everyone was afraid until finally, after much coaxing, Bat, who occupied the humblest seat near the door, consented to go.

At the home of Winter Thunder, whither they had gone, Black God forced Winter Thunder to listen to a proposal. When they sat down to talk it over there were Winter Thunder, Duck, Mudhen, Big White Duck, Beaver, Big Snake, and so many allies that they had to stand outside the house.

Eventually an agreement was concluded by the Dark Thunder and the Winter Thunder groups. As they were about to leave for the fight. Talking God told them to get in line. This time there were Monster Slayer, Child-of-the-Water, Changing Woman, Talking God, Big Fly, Pollen Boy, Cornbeetle Girl, the two Racing Gods and their grandmother, and the four Whirlwinds. All the Holy People, except Be'gotcidí, came: Cloud People, Water People, Fog People, Moss People. The same deities assembled at Dark Thunder's and at Winter Thunder's house. Talking God laid down a wide rainbow for Dark Thunder's party and took his place at its head. The other people stood on it behind him and Xa.ctcé.,ógan stood at the rear; the whole assembly was made invisible by a dark cloud. The organization of Winter Thunder's party on a white rainbow was exactly the same: Talking God was in front, Xa.ctcé.,ógan at the rear, and all were hidden by a white fog.

—from Gladys Reichard, *Navaho Religion*

Navaho Sandpainting-style Rug with Rainbow Guardian Border, c. 1920

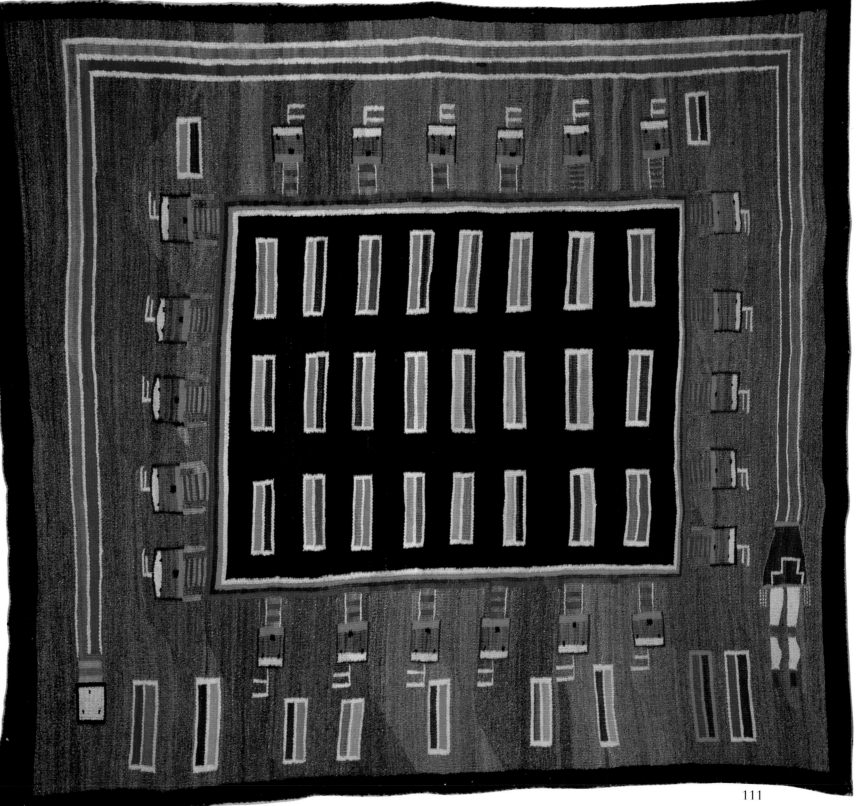

MORITZ VON SCHWIND, *THE RAINBOW*, c. 1860

Great Plains, said the rainbow was a clawmark of the red-bird. Some northeastern tribes believed the rainbow to be caused by the mist from the breaking surf on some great, distant body of water.

The Yukis of California believed the rainbow was the multi-colored garment of the Great Spirit who created all existence. A similar belief was found among the Cherokees, who—like the Samoyeds, a Siberian Mongol people who probably share common ancestors with the Native Americans—said it was the hem of the Sun God's coat.

David Thompson, the great explorer and mapper of Canada, related that the Indians accompanying him on one expedition joyfully exclaimed when they saw a magnificent rainbow after a great storm, "There is the mark of life, and we shall yet live."

The Mayan people of the Yucatan had a myth strikingly similar to that of Noah and the covenant. One of the so-called *Books of Chilam Balam* tells of vast destruction by an earthquake and a volcanic eruption. To the survivors of these simultaneous catastrophes a rainbow appeared as a sign that the destruction had ended and a new age begun. The Mayan goddess of the rainbow, Ixchel, was also associated with the moon, sexuality, childbirth and medicine.

AFTER THE DELUGE, God said to Noah, "I do set my Bow in the Cloud." That phrase is often taken to mean that God intended the rainbow to represent his immense bow of war, which he hung with its tips pointing downward, as a warrior lowers his bow when a battle is over.

Many cultures have thought of the rainbow as a divine weapon. We have already encountered the Italian word "arcobaleno" with that connotation. The Moslems of Iran call the rainbow the bow and arrow of Rustam, the sword of Ali, or the bow of Kuzah, which he hangs in the clouds when he has finished shooting his bolts of lightning. In Siberia, the rainbow is the thunder god's war bow.

The Assiniboins (also known as the Stoney tribe), a Sioux people of the Great Plains, believed that giants inhabited the world when it was very young. One day, the chief of these giants reached into the sky and grabbed the rainbow to use for hunting, but as soon as he seized it the bow became colorless. He then became so angry that he threw the bow against a mountain. It shattered and its pieces fell into a lake, in whose waters the colors of the shattered rainbow can sometimes be seen at sunrise.

BECAUSE ONE CAN NEVER reach the foot of a naturally occurring rainbow, there have been many superstitions about the rewards that will fall to anyone who manages this impossible feat. Throughout much of Europe, the treasure waiting at the foot of a rainbow is said to be a pot of gold, though some tales promise a handful of pearls or a single large and perfect pearl. Fairies are usually responsible for placing these rewards, but in Silesia, a region that has passed between Germany and Poland, angels put gold at the foot of a rainbow, and only a nude man can obtain the prize. One old Norwegian version held that the treasure would be a jug of *floedegroed*, a favorite concoction of cream and wheat, and that the fairies would considerately provide a golden spoon with which to eat it.

According to a superstition formerly found in France, Serbia, and Albania, anyone passing under the arch of a rainbow would undergo a change of gender. In Rumania it was said that anyone who crept to a rainbow's end on hands and knees and drank the water it touched would instantly change sex. A relatively modern Greek story told of a boy who was changed into a girl when he insulted a rainbow by jumping over it.

According to a Gypsy belief, eternal youth and beauty will be granted to anyone who manages to climb onto a rainbow at Whitsuntide, the week after Pentecost.

Some Europeans traditionally said whenever they saw a rainbow that God or Saint Peter had opened the door of Heaven, letting divine light shine down upon the Earth. Germans call the secondary rainbow the *Teufelsregenbogen*, or devil's rainbow. They say that such a pale bow is made by Satan in an unsuccessful attempt to outdo God.

The Chaga, a Bantu people of northern Tanzania, tell a tale about a man who stood where the rainbow touched the earth, praying for cattle. When no cattle appeared he cut the rainbow in two with his sword; half rose into the sky, and the other half disappeared into the earth, leaving a hole. It was said that people

PROVERBS AND SUPERSTITIONS

We say it is a sign of fair weather when there is a rainbow in the east, because when there is a rainbow in the east, it is alwaies already fair in the west for if it be cloudy there the rays of the sun will be hindered from coming thence to the opposite drops of rain.

—Jonathan Edwards, *Of the Rainbow*

Rainbow in the morning, sailors take warning;
Rainbow at night, sailors delight.
Rainbow to windward, foul fall the day;
Rainbow to leeward, damp runs away.

Rainbow in the morning,
Good man, be on your way.
Rainbow at night,
Good man, at home you should stay.

A rainbow at morn,
Put your hook in the corn.
A rainbow at eve,
Put your head in the sheave.

An old French superstition ordained that whenever a rainbow appeared, the children who saw it would all yell, "Arc-en-ciel! Arc-en-ciel!" Immediately, a boy who had not yet seen the rainbow was to pull a hair from his head and lay it on the palm of his left hand, oriented from wrist to fingertips. He was then to spit on the hair and say,

> *Rainbow, get out of my granary*
> *Or I'll cut you in half!*

If in the mornyng the raynebow appere, it signifieth moysture.
If in the evening it spend it self, fayr weather ensueth.

—L. Diggs, *Prognostication, 1555*

Medieval Germans believed that no rainbow would appear during the last forty years of the world's existence. Therefore whenever they saw a rainbow, they would take hope that the world would last at least another forty years:

> *So the rainbow appear,*
> *The world hath no fear,*
> *Until thereafter forty year.*

GIORGIO GHISI, *THE DREAM OF RAPHAEL* OR *ALLEGORY OF LIFE*, 1561

who climbed into the hole discovered an underground paradise, but lions drove them out and have driven away anyone who has gone down since.

THE RAINBOW is viewed by many cultures as an auspicious phenomenon, a symbol of hope, peace, and brotherhood. Because a rainbow most often follows a thunderstorm, the bow has been widely revered as a sign of reconciliation and peace after a conflict or an ordeal. But some cultures have traditionally viewed it as a sign of bad luck. Like comets and eclipses, the rainbow has often been regarded as ominous.

In Malaya, a partial arc ending in the water is taken to mean that some prince will soon die. The Roman natural historian Pliny said that the rainbow foretold a heavy winter or a war. The Yuki Indians believed that a rainbow brought dry weather; stretched across the sky, it prevented the rain from falling. The Arawak of South America believed the rainbow to be a fortunate sign if it was seen over the sea, but when it appeared on land it was an evil spirit searching for a victim. Dreaming of a rainbow has been widely regarded as an evil portent.

Iranian Moslems had a system of predictions based on the rainbow's appearance in the sections of the sky governed by each of the twelve signs of the zodiac. The appearance of a rainbow in the sign of the Ram in the east augured general good fortune; in the west, it meant famine. An appearance in Cancer in the east signaled prosperity; in the west, misfortune to the ruler. If the rainbow was seen in the sign of the Fish in the east, the priests would flourish; in the west, women would suffer. And so on.

These same people also thought of the rainbow as the reflection of light upon the magic mountain Qaf, which has seven peaks, each made of a jewel of a different color. The relative brightness of these colors in the rainbow was viewed as significant. If red predominated, it was thought that there would be war; green meant abundance; and yellow foreboded a terrible plague.

Among the Semang of Malaya, the places where a rainbow touches earth were considered unhealthy, contrary to a European superstition holding that land touched by a rainbow would become especially fertile. In his great twelve-volume work, *The Golden Bough; A Study in Magic and Religion*, the British anthropologist Sir James George Frazer relates that the people of the Indonesian island of Nias "tremble at the sight of a rainbow, because they think it is a net spread by a powerful spirit to catch their shadows," which is to say, their souls.

In Peru it was traditionally said that one must remain silent and place a hand over one's mouth when looking at a rainbow— not out of reverence, but out of fear that the rainbow would cause one's teeth to decay.

Many cultures have had injunctions against pointing at a rainbow. German folklore maintained that if one did so, one risked poking out the eye of an angel. Children of the Native American tribes of northern California were warned not to count the colors in a rainbow or to point at it. If they did so, the pointing finger would become crooked or drop off, or else they would become sick.

Since the rainbow is regarded as a manifestation of the "glory of the Lord," the Talmud declares that gazing at one is disrespectful and dangerous. Rabbi Joshua ben Levi thought that upon seeing a rainbow one should fall down upon the ground and touch one's face to the earth, but the Palestinian rabbis forbade this practice, which gave the impression that one was bowing down to the rainbow—as a heathen might do.

In the Talmud, the rainbow is considered a reminder of human wickedness and of the fact that only God's forebearance saves humankind from universal destruction.

A question long debated by theologians is whether or not rainbows had been seen before the Deluge. Since it would be heretical to suggest that God had created a natural phenomenon long after the Creation, the Talmud places the rainbow among the last things that He created at twilight on the sixth and final day of the Creation, before He declared Himself satisfied and consecrated the seventh day to rest. Regarding this controversy, we have, for instance, the following observations made by the eighteenth-century American Puritan minister Cotton Mather:

Tho' *Aben [Ibn] Ezra* thinks there was no Rainbow before the Flood, and many Christians also deny the *Antediluvian Existence* of it; yet very many will say with my learned *Helvetian*: They are very much mistaken. The Army of them who Engage on this side [in favor of the antediluvian rainbow] will have no less than a Great *Calvin* for a Standard-bearer. The Lord speaks not of the Rainbow as a *New Thing*; He seems to speak of it as a Thing that had already been placed in the *Clouds; I have set my Bow in the Cloud*. He invites us rather to Look on the *New Use* which He assigns unto it.

HUICHOL YARN PAINTING, MEXICO (DETAIL)

The Kumana of eastern Bolivia visualized the rainbow as a snake who threw stones at anyone who looked at him. Indeed, representations of the rainbow as a serpent were nearly universal, found among the tribes of North and South America, Africa, Indonesia, and Australia, as well as among many European peoples. Some European farmers have traditionally said that the rainbow is an evil serpent that comes down from the sky to quench its vast thirst in lakes and rivers. In Brittany, where the rainbow serpent was said to have the head of a bull and blazing eyes, folklore maintains that the rainbow prefers to remain invisible but can be seen when it becomes so thirsty that it is forced to descend to earth—at which point it drinks so deeply that it often swallows vast numbers of fish and frogs, which will subsequently rain down from the sky. In Slovakia it is said of a man attempting to quench a great thirst that he drinks like a rainbow. Such ideas of the rainbow go back to ancient times, for we read in Ovid's *Metamorphoses* that an angry Jove unleashed a violent thunderstorm upon the earth and then sent the rainbow, Iris, to "draw water from the teeming earth and feed it into the clouds again," so that the year's crops would all be ruined.

According to an Amazonian myth, the jungle birds—whose plumage was then dark and dull—joined forces with men to attack the evil rainbow serpent, which lived underwater and emerged periodically to swallow human beings. In a fierce battle, the giant snake was vanquished. As it expired, each bird miraculously acquired the colors of the part of the rainbow that it held in its beak. (According to an ancient Germanic creation myth, the rainbow is the bowl God used to hold his paints while coloring the birds.) The French anthropologist Claude Lévi-Strauss, in his book *The Raw and the Cooked: Introduction to a Science of Mythology*, explores at length the complicated concatenation of symbolic meanings involving the rainbow serpent through which the Amazonian Indians came to associate rainbows with poison and disease.

The rainbow serpent is of great importance in the mythology and rituals of the Australian aborigines. In *The Golden Bough* Frazer relates that among the Anula tribe of northern Australia a man who has the rain-bird for his totem is said to be able to make rain by catching a snake and holding it underwater in a certain pool. After killing the snake, he lays it on the ground beside a creek. "Then he makes an arched bundle of green stalks

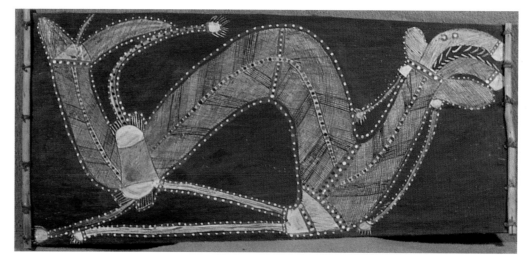

AUSTRALIAN BARK PAINTING OF A RAINBOW SERPENT

AMAZONIAN FEATHER HEADDRESS (URUBU PEOPLE)

119

in imitation of a rainbow, and sets it up over the snake. After that all he does is to sing over the snake and the mimic rainbow; sooner or later the rain will fall." According to Frazer, "The Kaitish tribe of Central Australia believe that the rainbow is the son of the rain, and with filial regard is always anxious to prevent his father from falling down. Hence if it appears in the sky at a time when rain is wanted, they 'sing' or enchant it in order to send it away."

A myth of the Luba people of Zaire relates that Nkongolo, the Rainbow King, became jealous of the strength and grace of his adopted son, Kalala Ilunga, and decided to kill him. Forewarned, Kalala returned to his real father, who gave him an army to attack Nkongolo. The latter fled to the mountains, but his twin sisters betrayed him to Kalala, who cut off his head. Nkongolo's spirit took the form of a serpent, which occasionally appears as a rainbow.

In the mythology of India, the supreme god Indra is armed with lightning arrows and a bow that appears in the sky as a rainbow when his anger has subsided. The elephant upon which he rides is Airāvata (whose name means "rainbow"), concerning whose birth there are two different myths. In one, as soon as Garuda, the "golden-winged sun-bird," emerged from its egg, Brahmā took the two halves of the shell and "sang over them seven holy melodies," giving birth to Airāvata. In the other version, Airāvata was a milk-white elephant born from the Milky Ocean after the gods and the titans had churned it for a thousand years.

Because the rainbow is insubstantial and ephemeral, it epitomizes the Buddhist conception of reality. Tibetan Buddhists therefore associate rainbows with the most enlightened gurus. They believe that through intense solitary meditation one may achieve the so-called "rainbow body"—a state of luminous awareness and bliss in which one is free from all desire. It is said that at death the body of one who has attained this state dissolves into light, the dissolution producing auras, beams, and circles of rainbow-colored light. The only physical traces left behind are the person's hair, fingernails, and toenails.

(ABOVE) TIBETAN THANGKA, *MAHAPARINIRVANA*, OR *DEATH OF THE BUDDHA*, 19TH CENTURY

(BELOW) THE RAINBOW-ENHALOED SHRINE IN THE LIBRARY OF THE PEMAYANGTSE MONASTERY, SIKKIM

RAGHUBIR SINGH, *RAINBOW OVER DAL LAKE, KASHMIR,* 1980

Twentieth Century

The rainbow bending in the sky,
Bedecked with sundry hues,
Is like the seat of God on high.

—George Gascoigne (c. 1535–1577), *Good Morrow*

GEORGE BELLOWS, *Well at Quevado*, 1917

ROBERT COZAD HENRI, *BOY AND RAINBOW*, 1902

The steadfast rainbow in the fast-moving, fast-hurrying hail-mist!

What a congregation of images and feelings, of fantastic permanence

amidst the rapid change of tempest—quietness the daughter of storm.

—Samuel Taylor Coleridge (1772–1834), excerpt from *Anima Poetae*

Victor Higgins, *Walking Rain* (or *Pablita Passes*), c. 1916–17

My heart leaps up when I behold

A rainbow in the sky:

So was it when my life began;

So is it now I am a man;

So be it when I shall grow old.

    Or let me die!

—William Wordsworth, 1802

CHARLES BURCHFIELD, *DREAM OF A STORM AT DAWN*, 1963–66

WASSILY KANDINSKY, *THE COSSACKS*, 1910–11

MAX BECKMANN, *THE RAINBOW*, 1942

131

I much question whether anyone who knows optics, however religious he may be, can feel in equal degree the pleasure of reverence which an unlettered peasant may feel at the sight of a rainbow.

—John Ruskin (1819–1900)

ROBERT DELAUNAY, *L'Air, le Fer, et l'Eau (Air, Iron, and Water)*, 1937

Know you that I have stumbled upon
    incredible Floridas,
With flowers and the eyes of human-
    skinned panthers,
And rainbows stretched out like brides.

—Arthur Rimbaud, *Le Bateau Ivre,* 1871

JOSEPH STELLA, *JOY OF LIVING,* 1938

Robert Delaunay, *Fleurs à l'arc-en-ciel*, c. 1925

Ay! let me like the ocean-Patriarch roam,
Or only know on land the Tartar's home!

…

Bound where thou wilt, my barb! or glide, my prow!
But be the star that guides the wanderer, Thou!
Thou, my Zuleika, share and bless my bark;
The Dove of peace and promise to mine ark!
Or, since that hope denied in worlds of strife,
Be thou the rainbow to the storms of life!

—George Noel Gordon, Lord Byron, *The Bride of Abydos*, 1813

ERICH HECKEL, *NORTH SEA NEAR OSTEND*, 1916

She saw the dun atmosphere over the blackened hills oppo-site, the dark blotches of houses, slate roofed and amor-phous, … And then, in the blowing clouds, she saw a band of faint iridescence colouring in faint colours a portion of the hill. And forgetting, startled, she looked for the hovering colour and saw a rainbow forming itself. In one place it gleamed fiercely, and, her heart anguished with hope, she sought the shadow of iris where the bow should be. Steadily the colour gathered, mysteriously, from nowhere, it took pres-ence upon itself, there was a faint, vast rainbow. The arc bended and strengthened itself till it arched indomitable, making great architecture of light and colour and the space of heaven, its pedestals luminous in the corruption of new houses on the low hill, its arch the top of heaven.

And the rainbow stood on the earth. She knew that the sordid people who crept hard-scaled and separate on the face of the world's corruption were living still, that the rainbow was arched in their blood and would quiver to life in their spirit, that they would cast off their horny covering of dis-integration, that new, clean, naked bodies would issue to a new germination, to a new growth, rising to the light and the wind and the clean rain of heaven. She saw in the rainbow the earth's new architecture, the old, brittle corruption of houses and factories swept away, the world built up in a living fabric of Truth, fitting to the over-arching heaven.

—D. H. Lawrence, excerpt from *The Rainbow*, 1915

I saw, almost frightened, a tremendous double rainbow of incredible sharpness and clarity.... Added to this was an intensity of color, the uncanny coppery glow of the sunset hour. The bow stretched across a deep violet sky, and in its midst appeared ghostly cloudheads in stratified rows.

—Lyonel Feininger, 1924

LYONEL FEININGER, *RAINBOW II*, 1928

Mine eyes, like clouds, were drizling raine

And as they thus did entertaine

The gentle Beams from Julia's sight

To mine eyes level'd opposite:

O Thind admir'd! there did appeare

A curious Rainbow smiling there;

Which was the Covenant, that she

No more wo'd drown mine eyes, or me.

—Robert Herrick (1561–1674), *The Rainbow: or Curious Covenant*

HANNAH HÖCH, *KUBUS*, 1926

*A*nd sometimes I remember days of old,

When fellowship seemed not so far to seek,

And all the world and I seemed much less cold,

And at the rainbow's foot lay surely gold,

And hope felt strong, and life itself not weak.

—Christina Rossetti (1830–1894), *The Thread of Life*

JIM DINE, *THE STUDIO LANDSCAPE*, 1963

Rainbow, stay,

Gleam upon gloom,

Bright as my dream,

Rainbow, stay!

—Alfred, Lord Tennyson, *Beckett, III,* 1884

If the world's a vale of tears,
Smile till rainbows span it!

—Lucy Larcom (1824-1893), *Three Old Saws*

RAINBOW SONGS

Over the Rainbow

Words by E.Y. Harburg, music by Harold Arlen

sung by Judy Garland in *The Wizard of Oz.*

I'm Always Chasing Rainbows

sung by Judy Garland

She's a Rainbow

sung by the Rolling Stones

Blinded by Rainbows

sung by the Rolling Stones

Swingin' on a Rainbow

sung by Frankie Avalon

Rainbow Road

sung by Joan Baez

Put a Rainbow in the Sky

sung by Mahalia Jackson

Rainbows Are Back in Style

sung by Dean Martin

Pocketful of Rainbows

sung by Elvis Presley

Ridin' the Rainbow

sung by Elvis Presley

There Should Have Been a Rainbow By Now

sung by Melanie

Rainbow on the River

sung by the Platters

Everybody Needs a Rainbow

sung by Bobby Wright

Lookin' for My Rainbow

sung by Canned Heat

JANET FISH, *DOUBLE RAINBOW*, 1996

Photographs

Just as silvery mist rises
Into the vast, empty firmament,
Will not the form of the lord guru
Appear in the immensity of all-pervading space?

Just as gentle rain slowly descends
Within the beautiful arc of a rainbow,
Will not the guru shower down profound teachings
Within a dome of five-colored light?

—Shabkar Tsongdruk Rangdrol (Tibetan lama, 1781–1851)

GALEN ROWELL, *RAINBOW OVER THE POTALA PALACE, LHASA, TIBET, 1981*

Vincent Parrella, *Niagara Rainbow (Maid of the Mist)*, 1996

TOM BEAN, *ICEBERG IN LE CONTE BAY WITH RAINBOW, TONGASS NATIONAL FOREST, ALASKA,* 1983   155

Tom Bean, *Rainbow and Cathedral Rock at Red Rock Crossing, Coconino National Forest, Arizona*, 1987

TOM BEAN, *Rainbow over Barn, Blue Earth County, East of Madelia, Minnesota,* 1994

WILLIAM EGGLESTON, *THE TRANSVAAL*, 1989

A Project by Fred Stern—The Rainbow Maker
"Millie's Rainbow," Baltimore Inner Habor, 1996
Photography by Max Wachsman

161

162   UPPER LEFT: *A RAINBOW FOR PEACE*, HUNTINGTON, LONG ISLAND, 1996.
OTHERS: *A RAINBOW FOR YOUTH*, LAS CRUCES, NEW MEXICO, 1996.

THE RAINBOW MAKER

Who, indeed, as Gerard Manley Hopkins asks in his poem, would make rainbows "by invention"? One answer to that question is Fred Stern, an environmental artist who engages the cooperation of local fire departments to create what he calls "Natural Man-Made Real Rainbows." Using the spray from fireboats or firetrucks, he has made his rainbows, some of them as large as 2,000 feet wide, in such cities as New York, Baltimore, Chicago, Miami, Salt Lake City, and Santa Fe. In 1992 he produced a series of rainbows for the Earth Summit in Rio de Janeiro, and three years later he presented a rainbow work entitled "Keshet Sheket," a Holocaust Memorial, at the opening of the Eutopia Festival in Potsdam, Germany. In the summer of 1996 Stern created a rainbow on New York City's East River so that it could be filmed by the Japanese national television network from an angle that made it appear to arc over the Secretariat building of the United Nations, as a symbol of world peace.

Stern has written, "I believe the rainbow is the true flag of our planet. So, whenever we see a rainbow, we are viewing the Earth displaying its colors and God's symbol for reconciliation. If more of us pledged allegiance to the Earth's flag rather than to dyed pieces of cloth, symbolizing national allegiances, our world would be a much better place to live. Remember this every time you see a rainbow, and feel free to display our planet's flag whenever you like by making your own rainbows."

At his website on the Internet (www.zianet.com/rainbow), Stern—who will soon be publishing a book to be entitled *The Rainbow Maker's Handbook*—provides detailed advice for others who wish to create rainbows in their communities. He makes this information freely available on the sole condition that it be used only for events in support of global unity and peace.

The basic prerequisites for a man-made rainbow are the same as those for one occurring spontaneously in nature. There must be bright sunlight behind the observers and a concentration of airborne drops of water in front of them. Since the sun must be neither too high nor too low above the horizon, the event must take place before mid-morning or after mid-afternoon. Stern recommends a time between one-and-a-half and three hours before sunset as the best. Since the sun will then be in the west, the spray of water must be toward the east.

If the town in which the rainbow is to be created is located on a river, Stern uses fireboats to generate the necessary spray. This is ideal, in that there are no problems regarding water supply or drainage. He uses a computer to analyze the site, to determine the optimal time for staging the event, and to calculate the required position of the fireboats relative to the observers. If fire hydrants and trucks must be used instead of boats, Stern has to provide for adequate drainage. In dry areas he compensates for the water usage by asking all of the observers to skip one shower or one toilet flush.

As for the greatest incalculable factor—the appearance or non-appearance of the sun—Stern remembers that his very first attempt to make a rainbow was a failure because the sun went behind a cloud ten minutes before the event was to take place. "After that I gave up being responsible for nature," says Stern. "I now request people coming to the event to bring the sun with them if they would like to see a rainbow." He has had no failures since that first one. For his efforts, Stern is rewarded by the expressions of joy and wonder on the faces of hundreds or even thousands of people, for almost everyone could say, as Wordsworth wrote, "My heart leaps up when I behold / A rainbow in the sky."

On a far less ambitious scale, Stern recommends looking for rainbows in the spray from sources such as a fountain, a waterfall, the ocean surf as it breaks on rocks, or the bow of a speedboat. He also tells children that during the summer they can make rainbows with a garden hose, using a misting nozzle or by holding a finger or thumb over a regular nozzle. "Go outside with your friends and run through the rainbow," he writes. "Rainbows can be quite refreshing. If you get in close enough not only will you get really wet, but you will get to see a full circle rainbow floating right in front of your face…. Give it a try. After you have made your own rainbows you will come to realize we are always surrounded by them. All you need to make them visible is the magic dust of a spray of water when the sun is out."

It was a hard thing to undo this knot,
The rainbow shines, but only in the thought
Of him that looks. Yet not in that alone,
For who makes rainbows by invention?
And many standing round a waterfall
See one bow each, yet not the same to all,
But each a hand's breadth further than the next.
The sun on falling waters writes the text
Which is yet in the eye or in the thought.
It was a hard thing to undo this knot.

—Gerard Manley Hopkins, untitled, 1880s

FRED STERN, *RAINBOW OVER THE UNITED NATIONS, NEW YORK*, 1996

CREDITS

All works are oil on canvas unless otherwise specified.

Amazonian headdress (Urubu people)
Cotinga, parrot, macaw, and pigeon feathers
Field Museum of Natural History, Chicago
119

Australian bark painting of a rainbow serpent, by the artist Namiriggi (Dangbon Group, Arnhem Land Plateau), mid-20th century
Edward L. Ruhe, University of Kansas
119

Bakoś, Józef
The Springtime Rainbow, 1923
Museum of Fine Arts, Museum of New Mexico, Santa Fe, gift of the artist in honor of Teresa Bakoś, 1974
145

Balthus (Balthazar Klossowski de Rola)
The Méditerranée's Cat, 1949
Private collection
109

Bean, Tom
A Glory Forms around the Photographer's Shadow, Canyon Rim, Colorado National Monument, 1994
Color photograph, © 1994 by Tom Bean
Courtesy of the photographer
33

Bean, Tom
Rainbow at Yaki Point on South Rim, Grand Canyon National Park, Arizona, 1990
Color photograph, © 1990 by Tom Bean
Courtesy of the photographer
84

Bean, Tom
Rainbow at Wukoki, Prehistoric Indian Ruin at Wupatki National Monument, Near Flagstaff, Arizona, 1990
Color photograph, © 1990 by Tom Bean
Courtesy of the photographer
108

Bean, Tom
Iceberg in Le Conte Bay with Rainbow, Tongass National Forest, Alaska, 1983
Color photograph, © 1983 by Tom Bean
Courtesy of the photographer
155

Bean, Tom
Rainbow and Cathedral Rock at Red Rock Crossing, Coconino National Forest, Arizona, 1987
Color photograph, © 1987 by Tom Bean
Courtesy of the photographer
158

Bean, Tom
Rainbow over Barn, Blue Earth County, East of Madelia, Minnesota, 1994
Color photograph, © 1994 by Tom Bean
Courtesy of the photographer
159

Beckmann, Max
The Rainbow, 1942
Private collection
131

Bellows, George
Fog Rainbow, 1913
Courtesy of Berry-Hill Galleries, New York
34

Bellows, George
Well at Quevado, 1917
Minnesota Museum of American Art, Saint Paul, Katherine G. Ordway Fund Purchase
124

Bierstadt, Albert
Rainbow over Jenny Lake, c. 1870
Private collection, courtesy of Sotheby's, New York
83

Blake, William
Los's Vision of Beulah from *Jerusalem*, 1808-1818
Hand-colored engraving
Yale Center for British Art, Yale University, New Haven, gift of Paul Mellon
105

Bolswert, Schelte Adams
Landscape with Rainbow, 1640s
Engraving after Rubens
Print Collection, The Miriam and Ira D. Wallach Division of Art, Prints, and Photographs; The New York Public Library Astor, Lenox, and Tilden Foundations
12

Brown, Ford Madox
Walton-on-the-Naze, 1859-60
City Art Gallery, Birmingham, England
90

Bruce, William Blair
The Rainbow, c. 1888
The Robert McLaughlin Gallery, Oshawa, Ontario, photo: Roy Hartwick
97

Burchfield, Charles
Dream of a Storm at Dawn, 1963-66
Thyssen-Bornemisza Museum, Madrid
Art Resource, New York
129

Church, Frederic Edwin
Rainy Season in the Tropics, 1866
Fine Arts Museums of San Francisco
Mildred Anna Williams Collection
73

Church, Frederic Edwin
Thunderstorm in the Alps, 1868
Oil and graphite on board
Cooper-Hewitt National Design Museum, Smithsonian Institution, gift of Louis P. Church; Art Resource, New York
Photo: Ken Pelka
74

Constable, John
London from Hampstead, with a Double Rainbow, 1831, watercolor
The Trustees of the British Museum, London
49

Constable, John
Salisbury Cathedral from the Meadows, 1831
National Gallery, London (on loan from the Dowager Lady Ashton)
59

Constable, John
The Grove, or Admiral's House, Hampstead, 1821-22
Victoria and Albert Museum, London
60

Constable, John
Landscape and Double Rainbow, 1812
Victoria and Albert Museum, London
61

Cropsey, Jasper Francis
The Narrows from Staten Island, 1866-68
© Amon Carter Museum, Fort Worth
Acquisition in memory of Richard Fargo Brown, Trustee, 1961-1972
86-87

Danby, Francis
The Shipwreck, n.d.
Wolverhampton Art Gallery, Staffordshire
Bridgeman/Art Resource, New York
63

Delaunay, Robert
L'Air, le Fer, et l'Eau (Air, Iron, and Water), 1937
Art Gallery of Ontario, Toronto
Samuel Zacks Collection, L & M Services B.V. 970611
133

Delaunay, Robert
Fleurs à l'arc-en-ciel, c. 1925
Private collection
135

Dine, Jim
The Studio Landscape, 1963
Property of A. Alfred Taubman, Bloomfield Hills, Michigan
146

Dossi, Dosso (Giovanni de' Lutteri)
Jupiter, Mercury, and Virtue, c. 1525
Kunsthistorisches Museum, Vienna
22

Duncanson, Robert Scott
Landscape with Rainbow, 1859
National Museum of American Art, Washington, D.C.; Art Resource, New York
78-79

Dürer, Albrecht
Melancholia I, 1514, engraving
Print Collection, The Miriam and Ira D. Wallach Division of Art, Prints, and Photographs; The New York Public Library Astor, Lenox, and Tilden Foundations
40

Eggleston, William
The Transvaal, 1989, color photograph
Art & Commerce, New York
160

Engel, Morris
Rainbow over New York, Seen Across Central Park, July 1996
Digitally-enhanced color photograph
© 1996 by Morris Engel
Courtesy of the photographer
156-57

Espingues, Evrard d'
The Rainbow, 1480
Page from manuscript of *Le Livre des Propriétés des Choses*, Ms. français 9140, fol. 205v, Bibliothèque Nationale, Paris
6

Fastolf Master
God Ends the Deluge, c. 1440-50
Miniature painting for a French book of hours, Ms. Auct. D. inf. 2.11, fol. 59v
Bodleian Library, Oxford University
20

Feininger, Lyonel
Rainbow II, 1928
Reynolda House, Museum of American Art, Winston-Salem, North Carolina
© 1997 Artists' Rights Society (ARS) New York / VG Bild-Kunst, Bonn
Photo: Jackson Smith
140-41

Fish, Janet
Double Rainbow, 1996
Courtesy of the artist
Photo: Beth Phillips Photography
151

Fisher, Alvan
The Great Horseshoe Fall, Niagara, 1820
National Museum of American Art, Washington, D.C.; Art Resource, New York
76

Flaxman, John
Iris with Venus and Mars, 1793
Engraving from a series for Homer's *Iliad*
106

Flötner, Peter
Several Ways of Light Refraction, 1535
Woodcut from an edition of *Optics* by Witelo (Vitellonis), printed in Nuremberg
44

Friedrich, Caspar David
Mountain Landscape with Rainbow, 1810
Museum Folkwang, Essen
53

Genthe, Arnold
Rainbow in the Grand Canyon, 1911
Autochrome (color photograph)
Library of Congress, Washington, D.C.
15

Genthe, Arnold
Rainbow over San Francisco, c. 1911
Autochrome (color photograph)
Library of Congress, Washington, D.C.
149

Ghisi, Giorgio
The Dream of Raphael or *Allegory of Life*, 1561
Engraving
Print Collection, The Miriam and Ira D. Wallach Division of Art, Prints, and Photographs; The New York Public Library Astor, Lenox, and Tilden Foundations
116

God's Pact with Noah
From the *Vienna Genesis*, 6th century A.D.
Codex theologicus graecus 31
National Library, Vienna
21

Grandville, Charles
L'Eventail d'Iris (The Fan of Iris), 1844
Hand-colored wood-engraving from *Un Autre Monde*
Print Collection, The Miriam and Ira D. Wallach Division of Art, Prints, and Photographs; The New York Public Library Astor, Lenox, and Tilden Foundations
18

Greenler, Robert
Infrared photographs of a rainbow, 1966 and 1970
Reproduced from his book *Rainbows, Halos, and Glories* (Cambridge University Press, 1980)
51

Gruenau, Douglas
Rainbow, Chaco Canyon, New Mexico, c. 1990
Color photograph, © 1997 by Douglas Gruenau, courtesy of the photographer
168

SELECTED BIBLIOGRAPHY

Boyer, Carl B. *The Rainbow: From Myth to
Mathematics.* New York, 1959. Reprint,
Princeton, 1987.

Graham, F. Lanier, ed. *The Rainbow Book,
being a collection of essays & illustrations devoted
to rainbows in particular & spectral sequences in
general, focusing on the meaning of color (physi-
cally and metaphysically) from ancient to modern
times.* Berkeley: Shambala [for] the Fine
Arts Museums of San Francisco, 1975.

Greenler, Robert. *Rainbows, Halos, and Glories.*
Cambridge University Press, 1980.

Landow, George P. "The Rainbow: A
Problematic Image." In *Nature and the
Victorian Imagination.* Edited by U.C.
Knoepflmacher and G.B. Tennyson.
Berkeley, 1977.

Mélusine (Paris, 1880s). During 1884 and
1885, this French periodical, devoted to
mythology and folklore, ran a regular
column about rainbow beliefs.

Menzel, Wolfgang. *Mythologische Forschungen
und Sammlungen.* Stuttgart & Tübingen,
1842. Section on the mythology of rainbows: pages 235-276.

Trilogie III: Regenbogen für eine bessere Welt.
Stuttgart: Württembergischer
Kunstverein, 1977.

Douglas Gruenau, *Rainbow, Chaco Canyon, New Mexico, c. 1990*